Computational Geometry on Surfaces

Computational Geometry on Surfaces

Performing Computational Geometry on the Cylinder, the Sphere, the Torus, and the Cone

by

Clara I. Grima

Department of Applied Mathematics (E.U.I.T.A.),
University of Seville, Seville, Spain

and

Alberto Márquez

Department of Applied Mathematics (F.I.E..),
University of Seville, Seville, Spain

KLUWER ACADEMIC PUBLISHERS
DORDRECHT / BOSTON / LONDON

A C.I.P. Catalogue record for this book is available from the Library of Congress.

ISBN 978-90-481-5908-6

Published by Kluwer Academic Publishers,
P.O. Box 17, 3300 AA Dordrecht, The Netherlands.

Sold and distributed in North, Central and South America
by Kluwer Academic Publishers,
101 Philip Drive, Norwell, MA 02061, U.S.A.

In all other countries, sold and distributed
by Kluwer Academic Publishers,
P.O. Box 322, 3300 AH Dordrecht, The Netherlands.

Printed on acid-free paper

A nuestras familias
To our families

Contents

Preface

In the last thirty years Computational Geometry has emerged as a new discipline from the field of design and analysis of algorithms. That discipline studies geometric problems from a computational point of view, and it has attracted enormous research interest. But that interest is mostly concerned with Euclidean Geometry (mainly the plane or Euclidean 3–dimensional space). Of course, there are some important reasons for this occurrence since the first applications and the bases of all developments are in the plane or in 3–dimensional space. But, we can find also some exceptions, and so Voronoi diagrams on the sphere, cylinder, the cone, and the torus have been considered previously, and there are many works on triangulations on the sphere and other surfaces.

The exceptions mentioned in the last paragraph have appeared to try to answer some questions which arise in the growing list of areas in which the results of Computational Geometry are applicable, since, in practice, many situations in those areas lead to problems of Computational Geometry on surfaces (probably the sphere and the cylinder are the most common examples). We can mention here some specific areas in which these situations happen as engineering, computer aided design, manufacturing, geographic information systems, operations research, robotics, computer graphics, solid modeling, etc. For instance, in geographic information systems and in operations research it is possible to consider worldwide questions which lead to problems in the sphere, in engineering or solid modeling are very common to deal with cases modeled by torus, cylinder or sphere. The cylinder is in general useful when we meet phenomena in which the same configuration appears in cycles. Finally, the arm of a robot does not describe, in general, an Euclidean space but a more complex algebraic surface, which in the simplest cases used to be one of the surfaces considered here.

As its title declares, this book is about Computational Geometry on Surfaces, but as its subtitle specifies, the material of this book is restricted to four very specific surfaces, the sphere, the cylinder, the cone, and the torus (in fact, this is not exactly true, in so far as we study some questions concerning more general surfaces, but we can say that more than ninety per cent of the book is devoted to the four surfaces mentioned). There are two main reasons for considering those surfaces. On one hand, they are the easiest surfaces after the plane, so naturally they must be the first to be considered when we try to travel beyond the plane. And, on the other hand, we think that restricting the material to those surfaces allows us to reach in an easier way the objective that we had in mind when we decided to start this work. So it is the intention of this book to demonstrate that classical problems of Computational Geometry can be solved when the input and output data are on surfaces other than the plane, but that planar techniques cannot be always adapted successfully and new techniques must be considered. In other words, we try to show here the flavor of Computational Geometry on surfaces.

Basically this book is conceived as a graduate text (in fact, its core is C.I. Grima's doctoral dissertation, although a lot of new material has been included), but we think that it can be useful to the professional in the applied fields mentioned above as well. Finally, it can be a guide for the researcher interested in Computational Geometry 'out of the plane', he or she can find here a sort of catalog of techniques in his/her discipline adapted to the surfaces considered here. In addition, some of the techniques and methods expound here can be adapted to other spaces that have not been treated directly but that share some common characteristics with the surfaces that we consider. Equally, we have tried to show not only how to obtain some results, but how it is impossible to obtain those results; in other words, which planar methods are not adaptable to our surfaces.

However, it must be pointed out that, as is common in this class of books, this book is not exactly a catalog of readily usable algorithms, but we focus mainly on the keys of the adaptation of planar algorithms to our surfaces.

Contents of the book

The three fundamental structures in Computational Geometry will be covered: convex hulls, Voronoi diagrams, and triangulations. These structures will be considered in three different surfaces, each one of them with some special characteristics: the sphere; the cylinder; and the torus (and occasionally the cone). In addition, some other classical problems will be studied: width; diameter; stabbing line; visibility; etc.

We will start with a first chapter that shows some notations and preliminary concepts, especially those regarding the metric of our surfaces as well as some distinguished elements. After that chapter we will focus on one of the key concepts of the book, Euclidean position. It is easy to imagine that performing Computational Geometry on surfaces will be different from the plane only when the input of an algorithm is a set that is not very small compared to the curvature of the surface. In fact, all surfaces considered here have closed geodesics, and, roughly speaking, we will see throughout this book that if the diameter of the set is smaller than half of the length of those geodesics, then to compute any invariant on that set coincides with the computations needed if the input is on the plane, in this case, we will say that the set is in Euclidean position. So, throughout the book, we will center our study on the cases of sets in non-Euclidean position.

In the third chapter we will study the convex hull of a set on a surface. In particular, two different extensions of convexity to our surfaces will be considered, and we will construct their convex hulls in optimal time. Both extensions will be based in the set of geodesics of each surface. Moreover, we will analyze the expected time of our algorithms and we will see that they run in linear time, but the bad news is that in most cases the convex hull is too big to be useful as a preprocessing for many problems (width, diameter, etc.), thus other tools must be constructed. In some sense this is one of the main and surprising characteristics of Computational Geometry on Surfaces, since in the plane, convex hull computations appear almost everywhere in order to solve a huge variety of problems, and we learn that, generally, in surfaces, that computation is useless for most of those problems.

The fourth chapter is devoted to Voronoi diagrams. As we have said above, these structures have been considered previously by other authors. So in that chapter we will summarize the known results on the closest point and the farthest point Voronoi diagrams. In addition, we will complete some of those known results, studying some extensions and giving valid methods for computing those diagrams in the cases which have not been considered previously.

So far all structures (convex hulls and Voronoi diagrams) studied in this work have the same complexity as in the plane, but in this chapter we will present a specific generalized Voronoi diagram (the polar diagram) that is more complex in the cylinder than in the plane.

The next two chapters are devoted to solving some practical problems that shape well the methodology which must be applied when doing Computational Geometry on Surfaces. Thus in Chapter four we will consider the computation of some functionals covered by the general name of radii, such as the width of a convex set or the diameter of a point set. And in the fifth chapter we treat the stabbing line of a segment set, and some visibility problems. In all cases we will show that the solutions given in the plane are not valid anymore (for instance, as we have mentioned above, the convex hull is not a useful tool for computing the diameter of a point set), but, in spite of that, optimal solutions can be found. This is an important part because it shows, probably better than any other part, the flavor of the field.

The last chapter introduces triangulations either of a point set or of a polygon. As in other structures studied before, some important differences appear in this subject. For instance, we can give two possible definitions of triangulations, which are equivalent in the planar case (a maximal subdivision and a triangular subdivision), and it is not clear whether both definitions agree outside the plane (we will see that both are equivalent in the case of the cylinder or the sphere but not in the case of the torus). The other three problems studied in this chapter are: what is the domain defined by a set of sites when we perform a triangulation?; is it possible to go from a triangulation to another using diagonal flips?; and, how can we obtain optimal triangulations? we will especially study the connectivity of the graph of triangulations of polygons on surfaces, seeing that, in general, but with some very remarkable exceptions, that graph is not connected.

Acknowledgments

It would be impossible to thank individually all our colleagues who have contributed to this book. We are grateful to all of them, even if we cannot list all their names here. However, we would like to explicitly thank (in alphabetical order) José Cáceres, Javier Cobos, Carmen Cortés, Juan C. Dana, Ángeles Garrido, Ferrán Hurtado, Felipe Mateos, Atsuhiro Nakamoto, Lidia Ortega, Joserra Portillo, Francisco Santos and Jesús Valenzuela. Without their contributions this book would never appear at the present time. We are also grateful to Dr. Liesbeth Mol and all the staff from Kluwer Academic Publisher for their support in all stages of the creation of this book. Last, but not least, our thanks go to our families for their support and love during the hours that have led to this finished product.

Chapter 1

PRELIMINARIES

Obviously, in a book like this the reader is usually familiarized with concepts and (some) results of Computational Geometry. In any case, we will try to make this book as self-contained as possible. So we introduce in this chapter some terminologies and notations that will be used along the book.

1. INTRODUCTION

Since the main subject of this book is Computational Geometry, it seems necessary to cite some classical texts on this discipline which constitute basic reference works on the objectives and methods used here: amongst them we can cite several titles such as: *Computational Geometry* [Preparata and Shamos, 1985], *Computational Geometry in C* [O'Rourke, 1994] and *Algorithms in Combinatorial Geometry* [Edelsbrunner, 1987]. In this way we must cite as well the recent books [Goodman and O'Rourke, 1997], [Sack and Urrutia, 2000], [de Berg et al., 1997] and [Boissonnat and Yvinec, 1998]. Regarding Voronoi diagrams, the obligatory references are the books [Okabe et al., 1992] and [Klein, 1989]. Concerning the design and analysis of algorithms, the texts [Aho et al., 1974] and [Knuth, 1973, Knuth, 1976] can be checked. A good reference on the metrics used in this book with a good presentation of other topics such as orbifolds, is the book by Nikulin and Shafarevich [Nikulin and Shafarevich, 1987]. Finally, regarding basic Differential Geometry the reader can refer to the book [do Carmo, 1992].

The model of computation used in this book is the real RAM model [Aho et al., 1974, Boissonnat and Yvinec, 1998], where each memory unit can

1

hold the representation of a real number, and accessing a memory location takes a constant time. With this model our computer works on real numbers of arbitrary precision for the same cost, so we can ignore all the problems related to numerical accuracy. The four elementary operations in this model are: comparison of two numbers, the four arithmetical operations, all the usual mathematical functions, and the integer part computation.

In this chapter, firstly we introduce some of the terminology used in the book, and we will summarize some of the properties of the surfaces that will be considered through the book. We will make special emphasis on those properties related to the family of geodesics, and on some distinguished elements which will be used through the book.

2.   NOTATIONS AND TERMINOLOGY

This section contains some details about the terminology used in the book, but it makes no pretence at providing formal definitions, just refreshing well known notions and introducing the adopted notation.

2.1   THE CYLINDER

At first glance working on the cylinder appears difficult in practice, but if we notice that a cylinder can be 'unrolled', this work becomes easier. So to study the geometry of the cylinder we will make use of a representation of it by unrolling it, or developing it onto the plane.

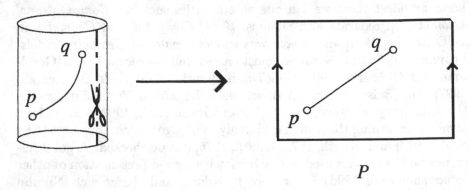

Figure 1.1.   To deal with the cylinder we will consider the representation of it obtained by unrolling it.

In order to 'walk' on this surface we will consider the planar representation of it as an infinite vertical tile a strip bounded by two parallel lines, $\mathcal{P}$, the two sides of which are identified. We can represent the geodesics in this surface using the tiling $\mathcal{T}$ defined by considering infinite copies of $\mathcal{P}$, see Figure 1.2; using this representation the geodesics can be treated as straight lines.

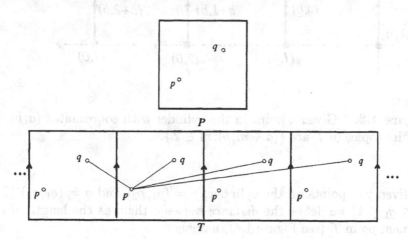

Figure 1.2. The infinity of geodesics joining p and q on the cylinder are represented in $\mathcal{T}$ by the straight lines joining p and the infinite copies of q.

A reference coordinate system is assumed in $\mathcal{T}$, as Figure 1.3 shows. For the sake of simplicity, and without lost of generality, we can assume that the distance between the two sides of $\mathcal{P}$ is 1. So given a point with coordinates (a, b) in this reference system, its infinite copies in $\mathcal{T}$ will be all the points belonging to the set $\{(a + m, b); m \in \mathbf{Z}\}$.

With this coordinate system the *parallels* or *great circles* of the cylinder will be the sets $\{(x, y_0)M; x \in \mathbf{R}\}$; and $\{(x_0, y); y \in \mathbf{R}\}$ will be the *generatrices* or *meridians* on this surface. We will say that two generatrices are *opposite generatrices* when the distance between them is $1/2$. We define the *segment* defined by two points p and q as the shortest geodesic in the cylinder joining them. When two points are on opposite generatrices (in this case both points will be called *opposite points*) the segment joining them is not unique. When we need to ensure the uniqueness of the segment joining two points in a set, we will consider as degenerate the cases in which there are two points on opposite generatrices, and to deal with these cases one can use the *S.o.S.* technique of Edelsbrunner [Edelsbrunner, 1987].

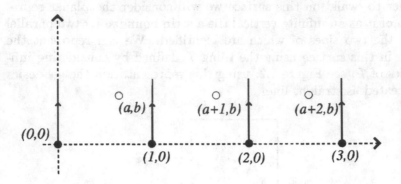

Figure 1.3. Given a point in the cylinder with coordinates (a, b), its infinite copies in $\mathcal{T}$ are $\{(a + m, b); m \in \mathbf{Z}\}$.

Given two points on the cylinder, $p = (p_1, p_2)$ and $q = (q_1, q_2)$ $(0 \leq p_1 \leq q_1 \leq 1)$ we define the *distance* between them as the length of the segment pq in $\mathcal{T}$ (see Figure 1.4), namely

$$d(p, q) = \begin{cases} \sqrt{(p_1 - q_1)^2 + (p_2 - q_2)^2} & \text{if } q_1 - p_1 \leq 1/2 \\ \sqrt{(p_1 + 1 - q_1)^2 + (p_2 - q_2)^2} & \text{if } q_1 - p_1 > 1/2 \end{cases}$$

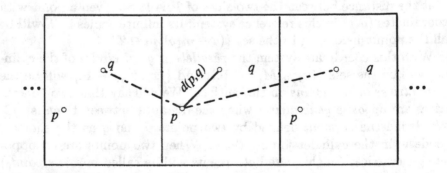

Figure 1.4. The distance between p and q is given by the length of the shortest geodesic joining them.

Finally, given a set $S = \{p_1, p_2, \ldots, p_N\}$ of sites $p_i = (x_i, y_i)$ on the cylinder such that $y_1 \leq y_2 \leq \cdots \leq y_N$, we define the *h-top of* S as the parallel $\{(x, y_N); x \in \mathbf{R}\}$ if $y_{N-1} = y_N$, or as the point $p_N = (x_N, y_N)$ otherwise; and, respectively, the *h-bottom of* S as the parallel $\{(x, y_1); x \in \mathbf{R}\}$ if $y_1 = y_2$, or as $p_1 = (x_1, y_1)$ otherwise. The *m-top* of S will be the smallest arc containing all points in S with y-coordinate equal to y_N if this arc is smaller than $1/2$, or $\{(x, y_N); x \in \mathbf{R}\}$ otherwise; and the *m-bottom* will be the smallest arc containing all points in S with y-coordinate equal to y_1 if this arc is smaller than $1/2$, or $\{(x, y_1); x \in \mathbf{R}\}$ otherwise.

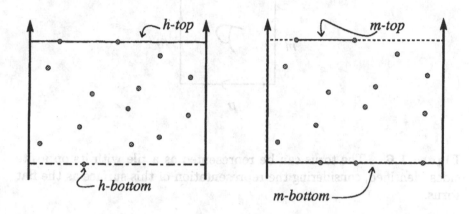

Figure 1.5. The *h*-top and the *h*-bottom are either whole parallels or single points; the *m*-top and the *m*-bottom are either whole parallels or arcs of them.

2.2   THE TORUS

In the case of the torus we can use, to deal with it, the well known representation of this surface in the *flat torus* [do Carmo, 1976]. With this representation the torus can be considered as a rectangle, or tile, $\mathcal{P}$ the opposite sides of which are identified in the same direction, see Figure 1.6.

As in the case of the cylinder, using this representation the geodesics can be treated as straight lines.

In order to work on the torus, we will consider the tiling $\mathcal{T}$ defined by the tile $\mathcal{P}$ by integer horizontal and vertical translations, see Figure 1.8.

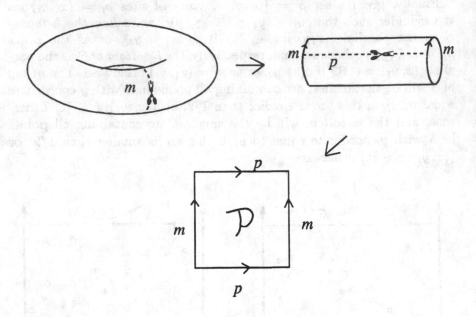

Figure 1.6. The torus can be represented as a tile with its opposite sides identified, considering the representation of this surface as the flat torus.

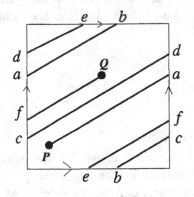

Figure 1.7.   A geodesic in the torus joining P and Q.

As in the cylinder, a coordinate reference system is assumed in $\mathcal{T}$, as Figure 1.9 shows.

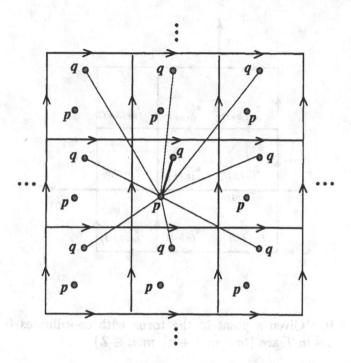

Figure 1.8. Considering the tiling defined by infinite copies of the flat torus, it will be easier to study the behavior of the geodesics in the torus.

With this reference system, the *parallels* of the torus are the straight lines $\{(x, y_0); x \in \mathbf{R}\}$; and the *meridians* are $\{(x_0, y); y \in \mathbf{R}\}$. We will say that the parallels $\{(x, y_1); x \in \mathbf{R}\}$ and $\{(x, y_2); x \in \mathbf{R}\}$ are *opposite parallels* in the torus when $|y_1 - y_2| = 1/2$; and, respectively, $\{(x_1, y); y \in \mathbf{R}\}$ and $\{(x_2, y); y \in \mathbf{R}\}$ are said to be *opposite meridians* if $|x_1 - x_2| = 1/2$. The square defined by two opposite parallels and two opposite meridians is called a *quadrant* of the torus.

Given two points p and q on the torus we define the distance between them, $d(p, q)$, as the length of the shortest geodesic joining them, and we call this geodesic the *segment* between p and q. This geodesic exists in the torus by virtue of the Hopf Rinow theorem [Hopf and Rinow, 1931], but it might be not unique, as happened on the cylinder, if p and q are on opposite parallels or opposite meridians. In any case, if we need to ensure the uniqueness of such a geodesic we will consider these cases as degenerate, and we can use some technique (*S.o.S.* of Edelsbrunner [Edelsbrunner, 1987], for instance) to manage them. In

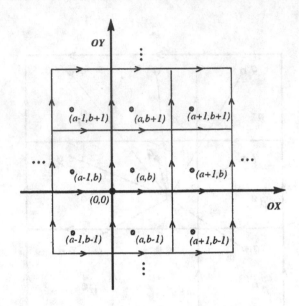

Figure 1.9. Given a point in the torus with coordinates (a, b), its infinite copies in $\mathcal{T}$ are $\{(a + m, b + n); m, n \in \mathbf{Z}\}$

concrete terms, if $p = (p_1, p_2)$ and $q = (q_1, q_2)$ are the coordinates of p and q in the torus, $d(p, q)$ is given by

$$d(p, q) = \min\{\sqrt{(p_1 - q_1)^2 + (p_2 - q_2)^2}, \sqrt{(p_1 - q_1 + 1)^2 + (p_2 - q_2)^2},$$
$$\sqrt{(p_1 - q_1)^2 + (p_2 - q_2 + 1)^2}, \sqrt{(p_1 - q_1 - 1)^2 + (p_2 - q_2)^2},$$
$$\sqrt{(p_1 - q_1)^2 + (p_2 - q_2 - 1)^2}, \sqrt{(p_1 - q_1 - 1)^2 + (p_2 - q_2 - 1)^2},$$
$$\sqrt{(p_1 - q_1 + 1)^2 + (p_2 - q_2 - 1)^2}, \sqrt{(p_1 - q_1 - 1)^2 + (p_2 - q_2 + 1)^2},$$
$$\sqrt{(p_1 - q_1 + 1)^2 + (p_2 - q_2 + 1)^2}\}$$

2.3   THE SPHERE

In the case of the sphere we can not use any isometric planar representation of it, so we must 'walk' on the whole surface. The geodesics of the sphere are the *meridians*, or *great circles*, defined by the intersection of the sphere with a plane containing its center.

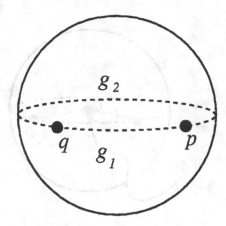

Figure 1.10. Given two points on the sphere p and q, there are two geodesics in this surface joining them.

In general, given two points p and q on the sphere a unique meridian is defined, and thus two unique geodesics joining them; the shortest one is called the *segment* joining p and q, and the *distance* between these points will be the length of this segment. When we talk about length on the sphere we will use radians, in such a way that the distance between p and q is the measure of the smaller angle defined in the plane Opq by the segments Op and Oq, being O the center of the sphere, see Figure 1.11. In the case that the two geodesics joining two points p and q on the sphere have the same length, π, we will say that p and q are *opposite* or *antipodal points*.

Given a meridian m of the sphere we will call the *poles* associated with m to the intersection points of the sphere with the straight line passing through O and orthogonal to the plane defined by m. Any meridian m_1 on the sphere has two common points with any other meridian m_2, say P_N and P_S; the tangents to the meridians in P_N (or in P_S) define four angles, such that opposite angles are equal; we call the *angle defined by m_1 and m_2* the smaller one. Moreover, these two meridians m_1 and m_2 define four regions in the sphere; we will call the *lune* associated with m_1 and m_2 the set defined for the two regions with less area; *the angle of the lune* will be the angle defined by the meridians. The points P_N and P_S will be called respectively *north pole* and *south pole*.

A *parallel* of the sphere is the intersection of any plane with the surface; in general it is not a geodesic. Given a parallel p we define the *equator* associated with p the meridian corresponding to the plane par-

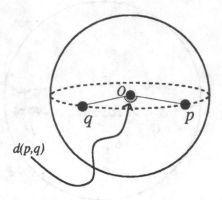

Figure 1.11.   The distance between p and q is measured by the angle defined by Op and Oq in the plane Opq.

allel to p passing through the center of the sphere; the poles associated with p will be the poles associated with the equator of p.

2.4   THE CONE

Again we can use a planar representation of our surface, just developing it by cutting by a straight line passing through the vertex of the cone, such a straight line will be called *a generatrix* of the cone. With this planar representation, geodesics will again be straight lines.

As in the cylinder, given two points in the cone it is possible to have two geodesics of minimum length, and, as in that surface, in the case of the cone we will say that these points are *opposite*. We call the *segment* joining p and q the shortest geodesic joining them (if this geodesic is not unique we have two segments). If we need to ensure the uniqueness of the segment between two points in the cone, we will consider the case with opposite points as a degenerate case. Remember that in the cylinder and in the torus, to manage with geodesics in these surfaces we consider a tiling $\mathcal{T}$ defined by infinite copies of the cylinder and of the torus, respectively. In this way, in the case of the cone, we will define a metric space $\mathcal{X}$ constructed as follows: consider three copies of the cone, say C_1, C_2 and C_3, and identify the right side of C_1 with the left side of C_2, and the right side of C_2 with the left side of C_3. Note that if the angle in the vertex of the cone is less or equal to $2\pi/3$, $\mathcal{X}$ is contained in

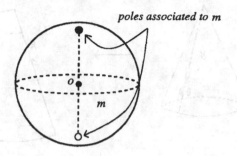

poles associated to m

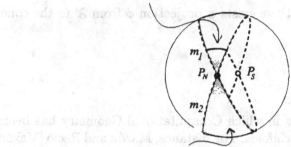

angle between m_1 and m_2

lune associated to m_1 and m_2

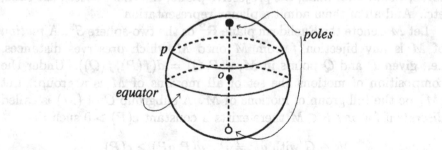

poles

equator

Figure 1.12. Several terminologies on the sphere introduced in the text.

the plane; in any case, $\mathcal{X}$ is locally isometric to the plane. Since any C_i

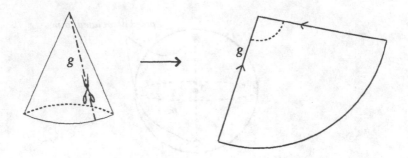

Figure 1.13.   Obtaining a planar representation of the cone.

is a develop of the cone, there exists a projection ϕ from $\mathcal{X}$ to the cone (see Figure 1.14).

3.   ORBIFOLDS

Another class of spaces in which Computational Geometry has been performed is the class of orbifolds. For instance, Mazón and Recio [Mazón, 1992, Mazón and Recio, 1997] have studied Voronoi diagrams for Euclidean and spherical orbifolds. This class of surfaces includes all the locally Euclidean and locally spherical surfaces such as the cylinder, the torus, the Möbius band, the projective plane, the Klein bottle, the cone, etc.. And all of them admit a planar representation.

Let M denote the Euclidean plane $\mathbf{R}^2$ or the two-sphere S^2. A *motion* of M is any bijection f from M onto M which preserves distances, i.e., given P and Q points in M, $d(P,Q) = d(f(P), f(Q))$. Under the composition of motions the set of all motions of M is a group. Let (M) be the full group of motions of M. A subgroup G of (M) is called *discrete* if for any $P \in M$ there exists a constant $c(P) > 0$ such that

$$\forall g \in G \text{ with } gP \neq P, \quad d(P, gP) \geq c(P),$$

where gP denotes the action of the motion g on point P. Note that the orbit of any point P under the action of a discrete group G, $GP = \{gP; g \in G\}$, is a closed and discrete subset of M. Any two points belonging to the same orbit will be called *equivalent* points. The quotient space M/G, whose points are the orbits of points of M under the action of G on M, inherits a natural metric from the metric on M. Given p and q orbits in M/G we define the distance $d(p,q)$ between them as the

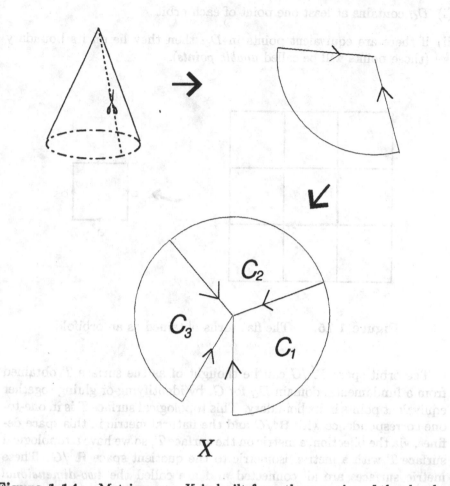

Figure 1.14.   Metric space X is built from three copies of the developed region of the cone.

distance between the sets p and q. With this distance it can be shown that M/G is a metric space, see [Nikulin and Shafarevich, 1987].

Note that in order to specify a point $p \in M/G$ we need only to know one point $P \in p$ in M, the remaining points in the orbit p are of the form gP for some $g \in G$. It is then useful, in order to handle the quotient space M/G, to determine some region in M which contains at least one point of each orbit and is as small as possible. If we consider the case $M = \mathbf{R}^2$, a *fundamental domain* for a discrete group G of Euclidean motions is a convex and closed subset D_G of the Euclidean plane with non-empty interior and satisfying:

(i) D_G contains at least one point of each orbit

(ii) if there are equivalent points in D_G then they lie on its boundary (those points will be called *double points*).

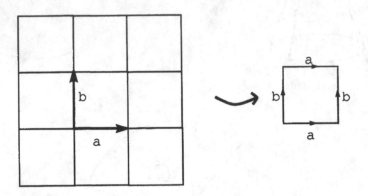

Figure 1.15.   The flat torus obtained as an orbifold.

The orbit space $\mathbf{R}^2/G$ can be thought of as the surface T obtained from a fundamental domain D_G for G, by identifying or gluing together equivalent points in its boundary. This topological surface T is in one-to-one correspondence with $\mathbf{R}^2/G$ and the natural metric in this space defines, via the bijection, a metric on the surface T, so we have a topological surface T with a metric, isometric to the quotient space $\mathbf{R}^2/G$. These metric surfaces are all connected and are called the *two-dimensional Euclidean orbifolds* ([Nikulin and Shafarevich, 1987]).   Observe that, for instance, the torus or the cone described previously are examples of orbifolds. In particular, the flat torus is the orbifold obtained from the group in which the two generators are two orthogonal translations in the plane. If the translations are not orthogonal then we obtain a torus with a metric other than that of the flat torus, which will play an important role in the last chapter.

4.   POINT LOCATION AND RANGE SEARCHING

There is not much to say about Point Location Problems or Range Searching on our surfaces. Of course, this kind of problem will be used

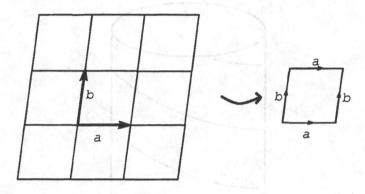

Figure 1.16. Another torus obtained as an orbifold that is not metrically equivalent to the flat torus.

in the same cases as in planar Computational Geometry. But notice that solving classical Point Location Problems is not a matter of any planar property but we just use that we have an appropriate coordinate system. As we have pointed out in previous sections, in our surfaces we do have that coordinate system, so the translation of the planar methods to solving Point Location problems is straightforward. In fact, we have described above planar representations (in Section 2.) of our surface (with the exception of the sphere) that suit perfectly for those purposes. Following these ideas several authors have proposed that Point Location on surfaces is just an adaptation of planar algorithms (see [de Berg et al., 1997]).

Nevertheless, some remarks must be made in order to clarify ideas. In the case of the cylinder, the torus, or the cone, the concept equivalent to straight lines is that of geodesics or helices (notice that we are considering the flat torus), but those lines do not divide the cylinder (or the torus or the cone) into two connected components, as Figure 1.17 shows, so some of the planar regions obtained by developing our surfaces must be identified in the Point Location Process. Moreover, two geodesics can intersect each other infinitely many times, so we must restrict our search to a band where the intersections are simple. In any case, the number of those intersections must be studied since this number determined the complexity (actually, the size of the input) for a Point Location problem.

In the case of the sphere exactly the opposite happens and so two geodesics split the sphere in four connected components. But we can

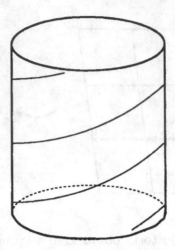

Figure 1.17. A helix does not split the cylinder into two parts.

restrict our search to each hemisphere, in which every pair of geodesics intersect in one point.

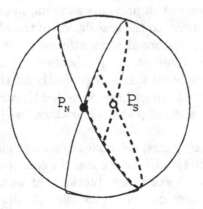

Figure 1.18. Two geodesics split the sphere into four connected components.

In some way Range Location problems are even easier to adapt, because they are based only in the coordinate system rather than in metric properties.

5.   NOTES AND COMMENTS

Every chapter of this book ends with a section called *Notes and Comments*. These sections are devoted to summarizing the results presented in the chapter and so as to present some open problems. Also we include some bibliographical references which could be useful for extending the study of the topics covered.

Chapter 2
EUCLIDEAN POSITION

If we try to compute the convex hull or a triangulation of a set of sites in a surface and all those sites are very close to each other, we can intuitively think that planar methods will be valid in this situation. This intuition has been used on several occasions by many authors, but sometimes it is not clear what 'very close to each other' means. In this chapter we will try to clarify this concept, introducing what we call Euclidean position, in such a way that if a set is in Euclidean position then we can work in that set as if it were in the plane.

1. INTRODUCTION

In most books on Computational Geometry there is mention of problems on surfaces, considering inputs such that planar methods are still valid for those data. Even in multiple applications of Computational Geometry it is assumed that if a given set is constrained to a small portion of a surface it presents a planar behavior. But we are not aware of a general framework for approaching the problem of deciding for which data planar methods are still valid. It is the aim of this chapter to introduce the concept of Euclidean position. Intuitively, if a set is in Euclidean position, all planar methods for solving problems in Computational Geometry are valid (or their adaptation will be very easy) for that set.

In some sense this is a book devoted to Computational Geometry on sets that are not in Euclidean position. It is in those sets that there are situations more interesting to analyze, and where there is more work to do; but, from a practical point of view, there exist many situations which can be solved by adapting planar algorithms in a straightforward way and we need to know when we are facing one of those situations.

The structure of this chapter is as follows. Firstly, we study Euclidean position in the surfaces considered throughout the book, taking into account mainly two problems. The first problem is deciding whether a set on a surface is in Euclidean position. And the second problem is to determine the probability of a set being in Euclidean position. Then we finish the chapter studying cylindrical position in the torus and some possible extensions of the concept to orbifolds.

2.   EUCLIDEAN POSITION

As we have said in the Introduction of this chapter, one of the key concepts of this book will be that of Euclidean position. If we have a set of sites $P = \{p_1, p_2, \ldots, p_N\}$ in any of our surfaces (cylinder, torus, cone, or sphere) we say that the set P is in *Euclidean position* in the corresponding surface when:

1. it is contained between two opposite generatrices of the cylinder or the cone;

2. it is contained in a quadrant of the torus; or

3. it is contained in a hemisphere of the sphere.

In some sense we will see that when a set is in Euclidean position its behavior will then be similar to the behavior of that set translated to the plane (this is to say, the convex hull will have the same shape on the surface as in the plane, and the same will happen with the significant portion of the Voronoi diagram, and so on). Thus, a relevant question in following chapters of this book will be to decide whether a set in one of our surfaces is in Euclidean position or not.

2.1   EUCLIDEAN POSITION ON THE CYLINDER AND THE CONE

Let us start with the cylinder and the cone in order to answer the question: whether or not a set in one of our surfaces is in Euclidean position. Note that deciding if P is in Euclidean position is equivalent to decide if, given a set of N points on a circle, the smallest arc covering them is less than π; it suffices to think in the orthogonal projections of the points in the cylinder on a circle, as in Figure 2.1.

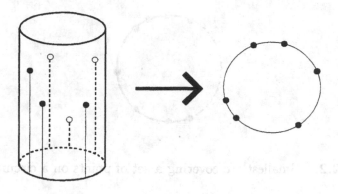

Figure 2.1. Deciding if a set is in Euclidean position in the cylinder.

It has been found that:

LEMMA 2.1 *[Sacristán, 1997] The computation of the smallest arc covering a set of N points on a circumference requires $\Omega(N \log N)$ operations.*

Proof: The argument of the author in [Sacristán, 1997] uses typical devices of lower bounds proofs for decision problems, that is, the transformation to the problem in question from another problem for which a lower bound has been established. In this case the chosen problem was:

MAXIMUM GAP PROBLEM. Given a set S of N real numbers $x_1, x_2, \ldots, x_N$, find the maximum difference between two consecutive members of S, where x_i and x_j are said to be consecutive if they are such in any permutation of $(x_1, x_2, \ldots, x_N)$ that has a natural ordering.

An $\Omega(N \log N)$ lower bound for the problem above had been established in [Lee and Wu, 1986].

Coming back to our problem, note that computing the smallest arc covering the set of points is equivalent to computing the largest empty arc, that is, the largest arc containing no points of the set, see Figure 2.2, this then justifies the choice of the *maximum gap problem*. $\square$

But we do not need to know exactly the smallest covering arc, it suffices to decide if this arc is less than π, and this decision problem can

Figure 2.2.   Smallest arc covering a set of points on a circumference

be solved in optimal time $O(N)$, as one can see in [Sacristán, 1997], using a separability criterion and linear programming techniques [Megiddo, 1983]. However, we will give here a simpler proof of this property.

LEMMA 2.2 *The problem of deciding if the smallest arc covering a set of N points on a circumference is less than π can be solved in optimal time $O(N)$. Moreover, if that arc is less than π then it can be found at the same computational cost. Therefore deciding whether a collection of points in the cylinder is in Euclidean position can be done in linear time.*

Proof:

   Let $\{v_1, v_2, \ldots, v_N\}$ be a set of points on a circumference. Consider a coordinate system of reference which origin O is the center of the circumference and the half-line Ov_1, for instance the positive X-axis, see Figure 2.3. Let $\mathrm{ang}_+(v_k)$ be the counterclockwise angle defined by Ov_k and Ov_1; and $\mathrm{ang}_-(v_k)$ the clockwise angle defined by these same points, with $k = 2, \ldots, N$.

   We call

$$v^+ = \max_{k=2,\ldots,N} \{\mathrm{ang}_+(v_k); 0 \le \mathrm{ang}_+(v_k) \le \pi\}$$

and

$$v^- = \max_{k=2,\ldots,N} \{\mathrm{ang}_-(v_k); 0 \le \mathrm{ang}_-(v_k) \le \pi\}.$$

Now it is clear that the smallest arc covering $\{v_1, v_2, \ldots, v_N\}$ is less than π if and only if $v^+ + v^- \le pi$. Computing v^+ and v^- requires $O(N)$ operations, since we need only to compute the maxima of two sets of, at most, N points.   □

   Our next question is: given a set $P = \{p_1, p_2, \ldots, p_N\}$ of sites chosen uniformly and independently at random from a cylinder, what is the

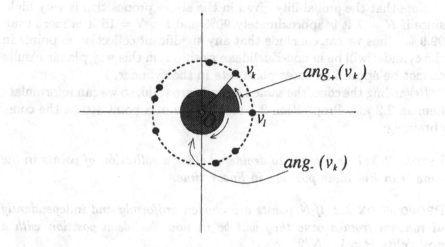

Figure 2.3. The coordinate reference system.

probability that P is in Euclidean position on this surface? Note that although elements can be chosen uniformly only from a set of bounded Lebesgue measure [Kendall and Moran, 1963] in our case, as only the abscissa will be significant and this coordinate describes a circle (and is therefore bounded), we do not need to specify in which bounded portion of the cylinder we are selecting our points (for any bounded portion our next Proposition will be valid)

PROPOSITION 2.1 *If N points are chosen uniformly and independently at random from a cylinder they will be in non-Euclidean position with a probability $P = 1 - N/2^{N-1}$.*

Proof: Let $P = \{p_1, p_2, \ldots, p_N\}$ be a set of N points on the cylinder, and let (x_i, y_i) be the coordinates of each p_i in the system of reference of this surface. If we call E the event 'P *is in Euclidean position*' we can express this event as $E = \bigcup_{i=1}^{N} E_i$, where E_i is the event 'P_i *is contained in the strip* $\{(x,y); x_i \le x \le x_1+1/2\}$' and $P_i = \{p_1, \ldots, p_{i-1}, p_{i+1}, \ldots, p_N\}$. Note that the events E_i are independent. The odds for each E_i are $1/2^{N-1}$, since every point of P_i is in the strip $\{(x,y); x_i \le x \le x_1+1/2\}$ with probability $1/2$. So, the odds for E are $\mathcal{P}(E) = \sum_{i=1}^{N} \mathcal{P}(E_i) = \dfrac{N}{2^{N-1}}$.

□

Note that the probability given in the above proposition is very high, since if $N = 7$ it is approximately 90% and for $N = 15$ it is more than 99.9%. Thus we can conclude that any significant collection of points in the cylinder will be in non-Euclidean position; in this way planar results cannot be applied for most point sets in the cylinder.

Regarding the cone, the same methods are valid, so we can reformulate Lemma 2.2 and Proposition 2.1 to the case of a point set on the cone, obtaining:

LEMMA 2.3 *It is possible to decide whether a collection of points in the cone is in Euclidean position in linear time.*

PROPOSITION 2.2 *If N points are chosen uniformly and independently at random from a cone they will be in non-Euclidean position with a probability $P = 1 - N/2^{N-1}$.*

2.2   EUCLIDEAN POSITION ON THE TORUS

Regarding the torus, using considerations similar to those in the case of the cylinder it is easy to design a linear algorithm to decide if a set of points is in Euclidean position or not. Basically, we have to apply the algorithm designed for the cylinder twice (once for the parallels and again for the meridians); so we have:

LEMMA 2.4 *It is possible to decide whether a collection of points in the torus is in Euclidean position in linear time.*

In fact, the idea of the algorithm can be used to solve the problem of the probability. Thus, we can prove that:

PROPOSITION 2.3 *If N points are chosen uniformly and independently at random from a torus they will be in Euclidean position with a probability $P = N^2/2^{2N-2}$.*

Proof: Let $P = \{p_1, p_2, \ldots, p_N\}$ be a set of N points on the torus, and let (x_i, y_i) be the coordinates of each p_i. If we call E the event 'P is in Euclidean position' we will have that $E = E_1 \cap E_2$, E_1 being the event 'P is contained between two opposite parallels' and E_2 the event 'P is contained between two opposite meridians'. As we did in the proof of Proposition 2.1, we can show that $\mathcal{P}(E_1) = \mathcal{P}(E_2) = N/2^{N-1}$. Since E_1 and E_2 are independent of each other, $\mathcal{P}(E) = N^2/2^{2N-2}$.   □

So by virtue of above proposition, in most cases the set of sites in the torus will be in non-Euclidean position; for $N = 10$ these odds are more than 99.96%.

2.3   EUCLIDEAN POSITION ON THE SPHERE

Using a separability criterion and linear programming techniques [Megiddo, 1983], it is possible to decide the same questions in the sphere, of whether the set of sites is in Euclidean position in linear time:

LEMMA 2.5   *[Sacristán, 1997] A set is in Euclidean position on the sphere if and only if that set is linearly separable from the center of the sphere, and this can be carried out in linear time (see Figure 2.4).*

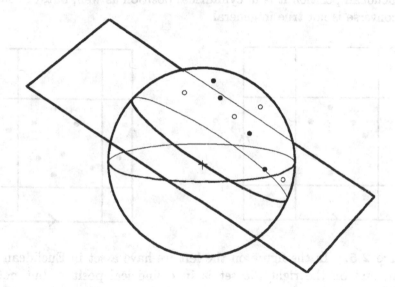

Figure 2.4.   A set in Euclidean position on the sphere can be separate from the center with a plane.

In any case, it is probably possible to develop a method applicable for this case avoiding the problem of deciding whether two sets are linearly separable. But we are not aware of the existence of such a method.

Regarding the probability of this proerty, we have as a consequence of a result of Wendel that:

PROPOSITION 2.4 *[Wendel, 1962] If N points are chosen uniformly and independently at random from a sphere, they will be in Euclidean position with a probability* $P = (N^2 - N + 2)/2^N$.

As in previous cases, the probability obtained in Proposition 2.4 is very low, so a set with 10 points will be in non-Euclidean position with a probability greater than 91%.

3.   CYLINDRICAL POSITION IN THE TORUS

In the case of the torus, in addition to Euclidean position we have another situation in which we can apply directly some methods obtained in this book for the cylinder. So we will say that a set of sites on this surface is in *cylindrical position* if the set of them is contained between two opposite parallels or two opposite meridians. Obviously, if a set is in Euclidean position it is in cylindrical position as well, but, of course, the converse is not true in general.

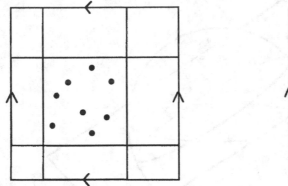
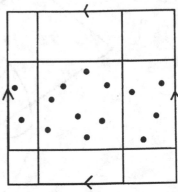

Figure 2.5.   In the figure on the left we have a set in Euclidean position, and on the right the set is in cylindrical position but not in Euclidean position.

One can decide if a set is in cylindrical position in linear time using the same arguments which we have used to decide whether a set of points is in Euclidean position in the cylinder. In that case we have to decide whether a set of points is between two opposite generatrices **and** between two opposite parallels. In this case we have to decide whether a set of points is between two opposite generatrices **or** between two opposite parallels. Moreover, using this same argument it is possible to prove:

PROPOSITION 2.5 *If N points are chosen uniformly and independently at random from a torus, they will be in cylindrical position with a probability* $P = \dfrac{N}{2^{N-2}}(1 - \dfrac{N}{2^N})$.

Proof: Let $P = \{p_1, p_2, \ldots, p_N\}$ be a set of N points on the torus, and let (x_i, y_i) be the coordinates of each p_i. If we call C the event 'P is in cylindrical position' we will have that $E = E_1 \cup E_2$, E_1 being the event 'P is contained between two opposite parallels' and E_2 the event 'P is contained between two opposite meridians'. As we did in the proof of Proposition 2.1, we can show that $\mathcal{P}(E_1) = \mathcal{P}(E_2) = N/2^{N-1}$. So $\mathcal{P}(E) = \mathcal{P}(E_1) + \mathcal{P}(E_2) - \mathcal{P}(E_1 \cap E_2)$, and the result holds. □

4. EUCLIDEAN POSITION IN ORBIFOLDS AND IN GENERAL SURFACES

As has been said before, arguments or algorithms on general surfaces that are valid in the planar case have been used in many applications , but it is not clear whether those arguments are still valid out of the plane. Thus an extension of the concept of Euclidean position to surfaces will be useful. Of course, this extension is not straightforward because some sets can preserve some planar conditions but not all, and, in many cases, it depends on the applications that we have in mind, but we can try here to make an approach to that problem.

As we have said above, orbifolds constitute a generalization of the surfaces considered in this book. In those spaces a good candidate for the concept of 'Euclidean position' can be given in the following way. If we call an *essential domain* the counter-image of the set obtained after deleting the double points of a fundamental domain, we say that a point set S is in *Euclidean position* if there exists an essential domain containing all the segments joining pairs of points of S. This definition agrees with that given in Section 2. in the cases of the cylinder, the sphere, the torus, or the cone. The problem is that a general algorithm for recognizing set of points in Euclidean position seems to be quadratic, and in order to obtain linear algorithms we need to study each class of orbifold separately.

Finally, there exists another alternative which is inspired by algebraic topology, that has the advantage that it can be applied to any surface: given a set P of points in a surface S the first step is to consider $G(P)$ defined by all segments joining points of P (in many ways, this process is almost the same that will be followed studying metrically convex hulls in the next chapter —see, for instance, Figures 3.6, 3.7 and 3.8—), and now we join every pair of points on $G(P)$ to get $T(P)$, then we say that P is in *Euclidean position* if $T(P)$ is simply connected.

Observe that with this last definition we can cover a wide range of cases, since when a set of points P is not in Euclidean position two options can happen:

- There exists an essential (i.e., homotopically non-null) closed curve in $T(P)$. For instance, in Figure 3.6 we can find an essential closed curve.

- There exists a hole in $T(P)$ (see Figure 2.6).

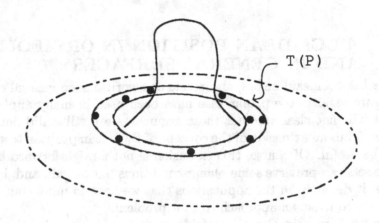

Figure 2.6. In this case $T(P)$ has a hole, and so the set is not in Euclidean position.

In some sense the first case has a topological nature and it covers all the cases considered above, but the second case is mainly metric and it is needed when the surface has a very high curvature in the proximity of the set P.

5.   NOTES AND COMMENTS

Although the main goal of this book is to perform Computational Geometry on the cylinder, the sphere, the torus, and the cone, if we consider another surface S probably the first question must be that of whether S is similar to one of those surfaces (topologically and metrically). We will see throughout this book that the behavior, and so the techniques to be applied, are not exactly the same in each case. But even in the case of having a surface similar, for instance, to the cylinder, we have to ask if the input is in 'Euclidean position', so an alternative definition of that concept must be found. A good candidate could be that considered in this chapter: if the surface have a system of closed geodesics, then we can say that a set of point is in Euclidean position if it lies between two opposite geodesics, where opposite geodesics must be defined with respect to a closed geodesic. Thus a pair of opposite geodesics could be two non-intersecting geodesics orthogonal to a closed surface in opposite vertices. On the other hand, and inspired in the definition of Euclidean position on orbifolds, we can think of an alternative based on choosing a specific geodesic and ensure that all segments joining pairs of points of a set in Euclidean position must avoid that geodesic.

Chapter 3

CONVEX HULL

Being a basic and natural concept, the convex hull has many applications as well as a rich mathematical theory behind it; moreover, the computation of planar convex hulls is one of the problems which has been most studied in Computational Geometry. Then it is natural, if we wish to carry out Computational Geometry on surfaces, to try to generalize these concepts (convexity and convex hull) to the surfaces we are considering here. When one tries to make that generalization every characterization of convex sets can be meaningful. In this chapter several extensions of this concept are considered in order to define and to compute the convex hull of a set of points on the cylinder, torus, sphere, and cone, obtaining algorithms which run in $O(N \log N)$ time in the worst case, but which are linear in almost all cases. This can sound as a good new but, unfortunately, when we can compute the convex hull in linear time we will have a hull which is too big and not very useful from a practical point of view.

1.   INTRODUCTION

The computation of the convex hull of a set has been studied extensively and has applications, for example, in pattern recognition [Akl and Toussaint, 1978] image processing [Rosenfeld, 1969] and stock cutting and allocation [Freeman, 1974, Sklansky, 1972, Freeman and Shapira, 1975] . Motion planning is another applied subject for which the convex hull provides an appropriate tool, for example in collision avoidance [O'Rourke, 1994], and a subfield of motion planning of considerable practical interest is planning the motion of an articulated robotic arm.

31

However, as we pointed out in the Preface of this book, in most cases the trajectory of an arm is not in the plane, but on a non-planar surface. Then the aim of this Chapter is to investigate the convex hull of a finite set of points on surfaces.

Several extensions of convexity to non-planar surfaces (or to non-Euclidean spaces) have been considered in the literature [Bielawski, 1987, Blanc, 1943, Busemann and Phadke, 1979, Pei, 1984]. Most of them are based on metric concepts, and, more concretely, on the family of geodesics of the surface. We will deal with those extensions in the case of the simplest surfaces, touching only two topics: *metric convexity* and *hyperconvexity*, but the interested reader can find more information on generalized convexity in the survey by Danzer et al. [Danzer et al., 1963].

1.1   SOME EXTENSIONS OF CONVEXITY

From a geometrical point of view convexity of a set $X \subset \mathbf{R}^n$ can be defined by requiring that the intersection of X with every straight line in $\mathbf{R}^n$ should be connected. If we replace the straight lines by some other family of curves then we are led to a concept of generalized convexity known as *hyperconvexity* [Danzer et al., 1963] (in Computational Geometry, a well known concept is that of orthogonal convexity, where the family of curves are two sets of orthogonal lines).

On the other hand, and according to Menger [Menger, 1928], a closed set $X \subset \mathbf{R}^n$ is called *metrically convex* if for every pair of points x and y in X there exists a point $z \in X \setminus \{x, y\}$ such that $d(x, z) + d(z, y) = d(x, y)$; in other words, if x and y are in X, then the segment joining both points must be contained in X.

In the remainder of this chapter we will characterize the convex hull on surfaces using the two concepts mentioned above. So Section 2 will be devoted to hyperconvexity and Section 3 to metrically convex sets. Also we include a section (Section 4) where we study the expected running time of the algorithms presented in the other sections. We will finish by solving a closely related problem, that of the minimum enclosing polygon of a point set.

2.   HYPERCONVEX HULL

As we have said before, it is possible to find several extensions of the notion of convexity taking into account some aspects of sets [Gruber,

1993, Mani-Levitska, 1993]; and every characterization of convex sets can be meaningful for generalizing this notion to non-planar surfaces. Regarding hyperconvexity in the case of non-planar surfaces, it seems natural to consider the geodesics as the family of curves, and, in this way, for a given surface S we can say that $P \subseteq S$ is *hyperconvex* if, given two points of P, all geodesics in S joining those points are contained in P. And, as usual, given a set P of points in S, the *hyperconvex hull* of P is the smallest hyperconvex set containing P. In general this definition is too stringent, as we shall see below

LEMMA 3.1 *The strip defined by two points of a hyperconvex set on the cylinder P is contained in P.*

Proof: Let x be a point in the strip defined by two points p and q of a hyperconvex set P, and let g and g' be two different geodesics joining p and q, hence g and g' must be contained in P. Let z and z' be the intersections of the circle containing x with g and g', respectively (see Figure 3.1). So x must be in one of the two geodesics joining z and z', then, since P is hyperconvex, x must be in P. □

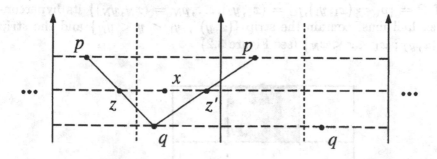

Figure 3.1. Point x must be contained in one of the two geodesics joining z and z'

According to the above lemma we can easily see that if P is a hyperconvex set in the cylinder then P is the union of an open strip defined by two parallels, with either one point or all the circle. Let $P = \{(x_1, y_1), (x_2, y_2), \ldots, (x_N, y_N)\}$ be a set of points on the cylinder with $y_1 \leq y_2 \leq \ldots \leq y_N$; let $O(P)$ be the open strip delimited by the maximal circles in the extreme points of P with respect to the ordinates $(O(P) = \{(x, y) : y_1 < y < y_N\})$; and define the *h-top* of P as the maximal circle $\{(x, y) : y = y_N\}$ if $y_{N-1} = y_N$ or the single point (x_N, y_N) otherwise (equivalently the *h-bottom*, see the Chapter 'Preliminaries'), so we obtain:

THEOREM 3.1 *The hyperconvex hull of a set P of two or more points on the cylinder is the union of $O(P)$ (defined above) and the h-top and the h-bottom of P.*

Proof: The proof of this result follows easily from Lemma 3.1. Since the hyperconvex hull of P must contain any strip defined by two of its points it must contain the strip $\{(x, y) ; y_1 < y < y_N\}$. Moreover, if $y_{N_1} = y_N$ the geodesics joining p_{N-1} and p_N must be included in the hull and they are contained in the circle $\{(x, y) ; y = y_N\}$ (which is the h-top of P in this case) so we must add this circle to the open strip in order to describe the hyperconvex hull of P. If $y_{N_1} < y_N$ the only point with ordinate y_N that we must add to the hull is p_N (which is the h-top of P now). The same proof works for the points with ordinate y_1.   □

In the case of the torus we can proceed analogously to the proof of Theorem 3.1 and prove the following theorem:

THEOREM 3.2 *The hyperconvex hull of a set P of two or more points on the torus is the whole torus.*

Proof: Using arguments similar to those used in the proof of Lemma 3.1, if $P = \{p_1 = (x_1, y_1), p_2 = (x_2, y_2), \ldots, p_N = (x_N, y_N)\}$ its hyperconvex hull must contain the strip $\{(x, y) ; y_1 < y < y_N\}$ and the strip $\{(x, y) ; x_1 < x < x_N\}$ (see Figure 3.2)

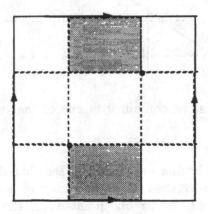

Figure 3.2.   The shadow strips in the figure are contained in the hyperconvex hull of the two points

Now, it suffices to join by geodesics the points in both strips, and we will cover the whole surface as a consequence.   □

When the set of points for which we wish to compute the hyperconvex hull is on the sphere, we will have that its hull is, as in the case of the torus, the whole surface. In order to see this property, consider two points p and q in a hyperconvex set of the surface; this set must contain the two geodesics joining p and q, that is, the meridian M defined for these points are completely contained in the set. Now we only need to draw all meridians joining the points of M pairwise in order to cover the sphere, see Figure 3.3

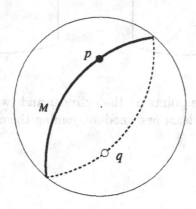

Figure 3.3. Given two points p and q in a hyperconvex set P on the sphere the great circle defined by them must be contained in P.

THEOREM 3.3 *The hyperconvex hull of a set P of two or more points on the sphere is the whole surface.*

In the case of a set of points in the cone, computing the hyperconvex hull is similar in spirit to computing the metrically convex hull, as we shall see in Section 3, so we refer the reader to that section.

Summarizing, we have that if we restrict our attention to the design of algorithms for computing the hyperconvex hull of a set of points on one of the our surfaces, these algorithms will run in linear or constant time, so one can feel lucky for this fact. Unfortunately, in those surfaces the hyperconvex hull is huge for many purposes, therefore it is convenient to use another definition of convexity.

In this way another possible choice could be to define a convex set as a set such that given two points in it there exists at least one geodesic joining them that is contained in the set. But this definition is even worse because it is not consistent, as we can see in Figure 3.4. In that figure we show three points and two sets containing them; both sets satisfy the condition given above but their intersection is even not connected.

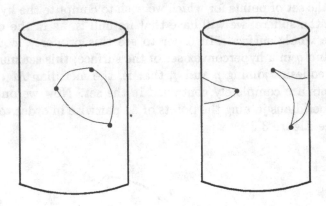

Figure 3.4. Three points on the cylinder and two different sets containing them and at least one geodesic joining them pairwise.

3.   METRICALLY CONVEX HULL

At this point we know that it is not good to consider all geodesics joining two points in order to define convexity on surfaces, and, even worse, to choose one at random; so it is time to consider the other topic mentioned in Section 1.1: *metric convexity*. As we have said in that section, according to Menger [Menger, 1928] a set $X \subset \mathbf{R}^n$ is called *metrically convex* if there exists for every pair x and y of points in X a point $z \in X \setminus \{x, y\}$ such that $d(x, z) + d(z, y) = d(x, y)$, in other words, if x and y are in X the segment joining both points must be contained in X. As we said in Section 2, following the Hopf–Rinow Theorem [Hopf and Rinow, 1931], we know that on the surfaces we are studying there always exists the segment joining two points, i.e., it is the geodesic of minimum length joining them. Actually, sometimes this segment is not unique; this happens when the points are opposite points on the surface. We will consider these cases as degenerate, and there exist several techniques for dealing with them, see, for instance, the *S.o.S. (simulation of simplicity)* of Edelsbrunner [Edelsbrunner, 1987].

Given a surface S we can agree with the definition given in [Menger, 1928] and say that $P \subseteq S$ is *metrically convex* (*m-convex* for short) if given two points of C the segment in S joining those points is contained in P. And, as usual, given a set P of points in S, the *metrically convex* hull (*m-convex hull*) of P, $C_m(P)$ is the smallest metrically convex set containing P.

Our purpose now is to characterize the m-convex hull of a set of points on the cylinder, the torus, the sphere, and the cone, and to give algorithms for computing it.

3.1 METRICALLY CONVEX HULL IN THE CYLINDER

We start with a set $P = \{p_1, p_2, \ldots, p_N\}$ of points on the cylinder and we can observe that if P is in Euclidean position, that is, it is contained between two opposite generatrices, the behavior of P is similar to a set of N points in the plane, all segments in the cylinder joining points of P are contained between these generatrices, (see Figure 3.5)

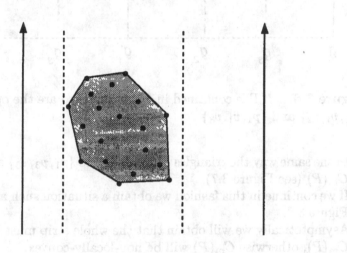

Figure 3.5. When the set of points in the cylinder is in Euclidean position the convex hull is similar to a set of points in the plane.

So, unless otherwise stated, we will consider that $P = \{p_1, p_2, \ldots, p_N\}$ is in non-Euclidean position on the cylinder and, under this hypothesis, we claim:

LEMMA 3.2 *The m-convex hull of three points in non-Euclidean position on the cylinder contains the open strip defined by those points.*

Proof: Let $P = \{v_1, v_2, v_3\}$ be three points in non-Euclidean position on the cylinder, and let Γ be the curve defined by the segments joining those points pairwise. *A priori* we know that S and Γ are contained in $C_m(P)$. Let g_i denote the generatrix containing v_i, and let g_i' be its opposite generatrix. If p_1 is the intersection of g_1' and the segment joining v_2 and v_3 it is easy to see that the triangles $\{p_1, v_1, v_2\}$ and $\{p_1, v_1, v_3\}$ are contained in $C_m(P)$ (see Figure 3.6).

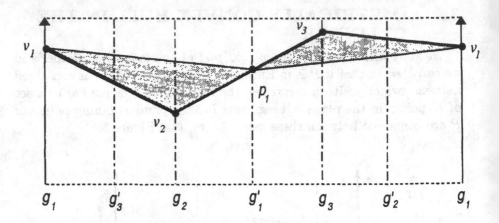

Figure 3.6. If Γ is contained in $C_m(P)$ then so are the open triangles $\{p_1, v_1, v_2\}$ and $\{p_1, v_1, v_3\}$

In the same way the triangles $\{p_1, p_2, v_2\}$ and $\{p_1, p_3, v_3\}$ are contained in $C_m(P)$ (see Figure 3.7).

If we continue in this fashion we obtain a situation such as that shown in Figure 3.8.

Asymptotically we will obtain that the whole strip must be contained in $C_m(P)$, otherwise $C_m(P)$ will be non-locally-convex.   □

We are now in a position to characterize the m-convex hull of a set of points on the cylinder as follows. Let $P = \{(x_1, y_1), (x_2, y_2), \ldots, (x_N, y_N)\}$ be a set of points in the cylinder with $y_1 \leq y_2 \leq \ldots \leq y_N$; $O(P)$ will be the open strip $O(P) = \{(x, y) : y_1 < y < y_N\}$; and define the *m-top* of P as the minimal arc containing all points of P with the ordinate y_N if that arc is shorter than a half of the circle, or as the whole maximal circle if that arc is greater than a half of the circle, or as the single point (x_N, y_N) otherwise (equivalently the *m-bottom*)

THEOREM 3.4 *The m-convex hull of a set of N points P in the cylinder is*

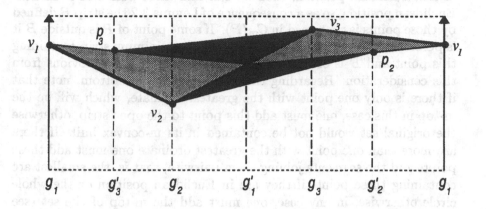

Figure 3.7. p_2 and p_3 are the intersection the segments joining p_1 and v_1 with $g'2$ and $g'3$ respectively

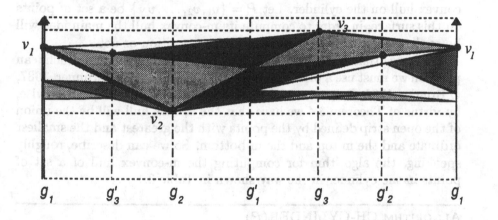

Figure 3.8.   All triangles appearing in this figure are contained in $C_m(P)$

1. *The convex hull of P in the plane if P is in Euclidean position.*

2. *The union of the open strip $O(P)$ and the m-top, and the m-bottom if P is not in Euclidean position.*

Proof: If P is in Euclidean position it is clear that the region in the cylinder corresponding to the convex hull of P in the plane is the m-convex hull of P in the cylinder. Assume that $P = \{v_1, v_2, \ldots, v_N\}$ is in non-Euclidean position, then we can find three points in P in non-Euclidean position, so as a consequence of Lemma 3.2 the strip B defined by those points is contained in $C_m(P)$. If some point of P is outside B it is possible to show, as in Lemma 3.2, that the minimum strip containing this point and B is contained in $C_m(P)$, so the result is obvious from this consideration. Regarding the m-top and the m-bottom, note that if there is only one point with the greatest ordinate, which will be the m-top in this case, one must add this point to the open strip, otherwise the original set would not be contained in its m-convex hull. If there are more than one point with the greatest ordinate one must add these points and the segments joining them pairwise, that is, the smallest arc containing these points if they are in Euclidean position or the whole circle otherwise; in any case, one must add the m-top of the set (see Figure 3.9). Similar arguments prove that the m-bottom must be added as well.

$\square$

Using Theorem 3.4 we can design an algorithm to compute the m-convex hull on the cylinder. Let $P = \{v_1, v_2, \ldots, v_N\}$ be a set of points on this surface; in order to compute its m-convex hull the main idea will be to determine whether P is in Euclidean position or not.

We know that if the set of points in the cylinder is in Euclidean position we must use a planar convex hull algorithm [Edelsbrunner, 1987, O'Rourke, 1994, Preparata and Shamos, 1985, Seidel, 1997]. Otherwise, by virtue of Theorem 3.4 we have that the convex hull will be the union of the open strip defined by the points with the greatest and the smallest ordinate and the m-top and the m-bottom. So we can describe, roughly speaking, the algorithm for computing the m-convex hull of a set of points in the cylinder in the way shown in Table 3.1.

ALGORITHM CH-CYLINDER(P)

INPUT: $P = \{v_1, v_2, \ldots, v_N\}$ *set of points on the cylinder.*

OUTPUT: *Metrically convex hull of P in the cylinder.*

STEP 1 *If P is in Euclidean position go to Step 3.*

STEP 2 *Compute the m-top and the m-bottom of P.(END)*

STEP 3 *Compute the convex hull of P in the plane.(END)*

Table 3.1. Algorithm for computing the metrically convex hull of a set on the cylinder

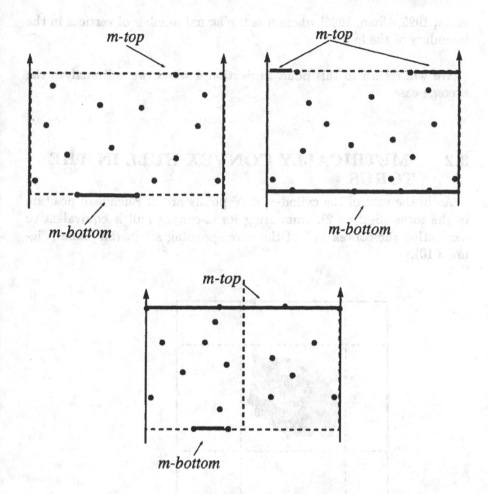

Figure 3.9. Some sets and their *m*-convex hulls.

Therefore we can summarize this study on the cylinder by the following theorem:

THEOREM 3.5 *Algorithm* CH-CYLINDER(P) *computes the m-convex hull of a set P of N points on the cylinder in optimal $O(N \log h)$ time, h being the number of points in the boundary of the hull.*

Proof: The result is straightforward since using Lemma 2.2 we have that Step 1 requires linear time; Step 2 just computes the maximum and the minimum of a set of N points, so it needs linear time too; and regarding Step 3 it is well known that optimal convex hull algorithms in the plane run in $O(N \log h)$ time [Seidel, 1997, Kirkpatrick and Seidel, 1986, Chan et al., 1995, Chan, 1995] where h is the actual number of vertices in the boundary of the hull.   □

We will return to this point in Section 4 where we will analyze the average case.

3.2   METRICALLY CONVEX HULL IN THE TORUS

As in the case of the cylinder, if N points are in Euclidean position in the torus (Section 2), computing its m-convex hull is equivalent to computing the convex hull of the corresponding set in the plane (Figure 3.10).

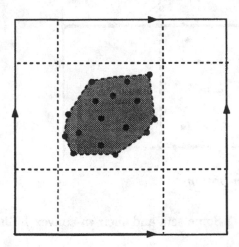

Figure 3.10.   When the points are in Euclidean position in the torus, its m-convex hull is similar to its convex hull in the plane

If the points are in *cylindrical position* (see Preliminaries) on the torus the m-convex hull of these points can be computed as being on the cylinder. If the set $P = \{v_1, v_2, \ldots, v_N\}$ is contained in the strip defined by two opposite meridians (respectively two parallels) of the torus, m_1 and m_2, it is straightforward to see that all segments joining points of P are also contained in this strip, so it suffices to consider the cylinder obtained by cutting the torus with m_1 and m_2, see Figure 3.11.

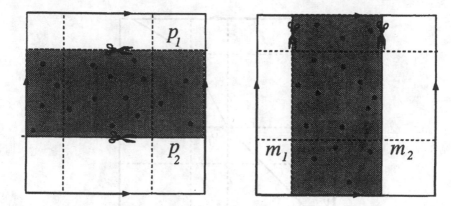

Figure 3.11. When the points on the torus are in cylindrical position one can consider these points as being on a cylinder by cutting the torus.

On the other hand, if the points are in non-cylindrical position we have:

LEMMA 3.3 *The m-convex hull of three points in non-cylindrical position on the torus is the whole torus.*

Proof: Let $P = \{v_1, v_2, v_3\}$ be a set in non-cylindrical position on the torus, without loss of generality we can consider v_1 in the center of the flat torus; let Γ be the curve defined by the segments joining the points of P pairwise; we call m_i and p_i respectively the meridian the parallel and containing v_i; and m_i' and p_i' will be their respective opposites, see Figure 3.12.

We know that at least P and Γ must be contained in the m-convex hull of P. So if we consider the cylinder defined by m_1 and m_1' containing v_2, and compute the hull of the points of Γ on this cylinder, using the results of the section above on this surface, we will have that this strip must be contained in the m-convex hull of P on the torus, Figure 3.13.

If we repeat this procedure on the cylinder defined by p_1 and p_1' containing v_3, we will have that this cylinder is also contained in the hull of P in the torus.

It suffices to join the points in both strips by segments in order to cover the whole torus.   □

According to the above considerations we claim:

THEOREM 3.6 *The m-convex hull of a set of N points P in the torus is:*

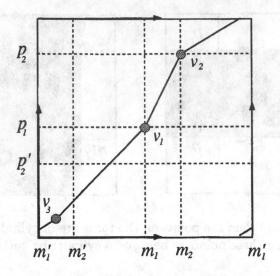

Figure 3.12.   Three points on the torus in non-cylindrical position.

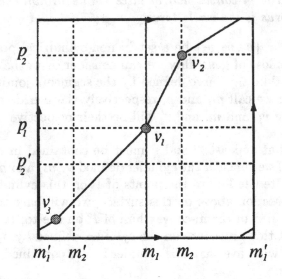

Figure 3.13.   The shaded strip is contained in the m-convex hull.

1. *The convex hull of P in the cylinder if P is in cylindrical position;*
   or

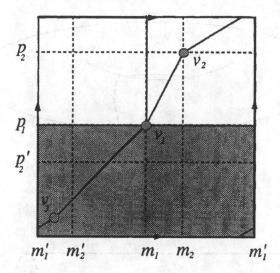

Figure 3.14.   The shaded horizontal strip is contained in the m-convex hull.

2. The whole torus, otherwise.

As in the case of the cylinder, the result above provides an algorithm for computing the m-convex hull on the torus of a set P, as follows:

1. Cut the torus by a meridian to obtain a cylinder, Figure 3.15. Then one can ask and answer in linear time, using Lemma 2.2, whether or not the points of P in this cylinder are in Euclidean position. If the answer is YES the points in the torus are in cylindrical position and we can use algorithm CH-CYLINDER(P). If the answer is NO go to next step.

2. Now cut the torus by a parallel and test whether P is in Euclidean position in the obtained cylinder. If the answer is YES P is contained between two opposite meridians, therefore it is in cylindrical position and one can use algorithm CH-CYLINDER(P) in order to compute its m-convex hull. If the answer is NO the m-convex hull of P is the whole surface, see Theorem 3.6.

We can therefore attempt a rough sketch of the algorithm as is described in Table 3.2.

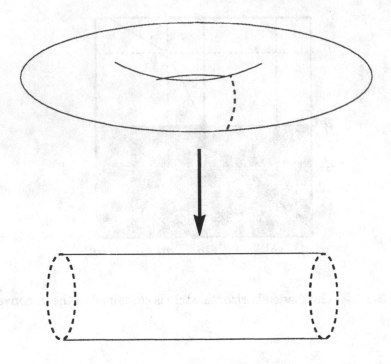

Figure 3.15.   Cutting the torus by a meridian we obtain a cylinder.

ALGORITHM CH-TORUS(P)

> INPUT: $P = \{v_1, v_2, \ldots, v_N\}$ *set of points on the torus.*
>
> OUTPUT: *Metrically convex hull of P in the torus.*
>> STEP 1 *If P is in cylindrical position go to Step 3.*
>> STEP 2 *The m-convex hull is the whole torus. (END)*
>> STEP 3 *Compute the convex hull of P in the cylinder.(END)*

Table 3.2.   Algorithm for computing the metrically convex hull of a set on the torus.

THEOREM 3.7 *Algorithm* CH-TORUS(P) *computes the m-convex hull of a set P of N points in the torus in optimal $O(N \log h)$ time, h being the number of points in the boundary of the convex hull.*

Proof: This result is straightforward since Step 1 requires linear time, see Lemma 2.2; Step 2 can be performed in constant time; and, regarding Step 3, convex hull algorithms in the cylinder run in optimal time $O(N \log h)$ where h is the actual number of vertices of the convex hull, as we have shown in Theorem 3.5. □

3.3 METRICALLY CONVEX HULL ON THE SPHERE

Our next task will be to characterize and to compute the metrically convex hull of a set of N points on the sphere. As we have stated in the second chapter, a set P on this surface is in Euclidean position if it is contained in a hemisphere. In this sense, computing the m-convex hull $C_m(P)$ on the sphere will be similar to doing so on the cylinder and on the torus: when P is in Euclidean position it suffices to adapt a planar algorithm, otherwise the m-convex hull is the whole sphere and so is too big for many purposes.

LEMMA 3.4 *The m-convex hull of four points in non-Euclidean position on the sphere is the whole surface.*

Proof: Let $P = \{v_1, v_2, v_3, v_4\}$ be a set of four points in non-Euclidean position on the sphere. Let C_{ij} be the great circle defined by v_i and v_j, and let γ_{ij} be the segment joining these points, so $\gamma_{ij} \subset C_{ij}$. If we consider, for instance, C_{12}, since P is in non-Euclidean position, v_3 and v_4 then lie in the two different hemispheres defined by C_{12}. The segment γ_{12} does not intersect γ_{34}, otherwise if we were to consider the hemisphere which pole is the intersection point of these segments, we would have that P is in Euclidean position, see Figure 3.16 (a).

We know that γ_{12} and γ_{34} are included in the hull of P and so are the geodesics $C_{12} \setminus \gamma_{12}$ and $C_{34} \setminus \gamma_{34}$. In fact, since the triangles $\{v_1, v_2, v_3\}$ and $\{v_1, v_2, v_4\}$ must be contained in the hull, and $C_{34} \setminus \gamma_{34}$ is included in those triangles; and the same holds for $C_{12} \setminus \gamma_{12}$ using the triangles $\{v_1, v_3, v_4\}$ and $\{v_2, v_3, v_4\}$, Figure 3.16 (b). So we have that C_{12} and C_{34} are contained in $C_m(P)$. It suffices to join the points on these two circles pairwise in order to cover the whole surface (see Figure 3.16 (c)), since the four lunes defined by those two geodesics have less than 180 degrees. □

So we have that if the set P of N points, $N \geq 4$, is in non-Euclidean position the m-convex hull of P will be all the sphere; but, what about

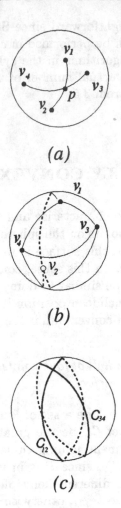

Figure 3.16. (a) If γ_{12} intersects γ_{34} the points $\{v_1, v_2, v_3, v_4\}$ are in Euclidean position, they are contained in the hemisphere the pole of which is p; (b) The triangles $\{v_1, v_2, v_3\}$ and $\{v_1, v_2, v_4\}$ must be contained in the m-convex hull of $\{v_1, v_2, v_3, v_4\}$, and the geodesic $C_{34} \setminus \gamma_{34}$ is contained in these two triangles; (c) If C_{12} and C_{34} are contained in the hull, it suffices to join, by segments, the points on both geodesics pairwise and to cover all the sphere, therefore the hull is the whole sphere.

the hull of a set in Euclidean position? One cannot apply directly a planar algorithm for computing the hull in the sphere, since we do not have a planar representation of this surface. But, actually it is very easy to describe a simple adaptation of Graham's algorithm, given originally for the plane, to the sphere:

Let $P = \{v_1, v_2, \ldots, v_N\}$ be the set of points on the sphere.

mbox

1. In order to obtain an internal point of the hull compute the centroid of any three non-collinear points of P [1]; let p be this centroid. Transform the coordinates of $\{v_1, v_2, \ldots, v_N\}$ to make p the north pole of the sphere, and sort the N points lexicographically by polar angle and distance from p, see Figure 3.17.

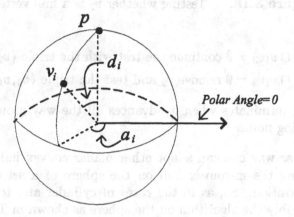

Figure 3.17.   If p is the north pole, a_i is the polar angle of v_i from p, and d_i is the distance of v_i from it.

Note that we do not compute the distance between every pair of points of P and, moreover the distance comparison is made only if two points have the same polar angle, that is, when they and p are collinear.

2. Compute one of the furthest points of p which is certainly a hull vertex. Start the scan from it around the ordered points.

3. Examine all triples of consecutive points in the circular order; for any triple $\{v_1, v_2, v_3\}$, test whether segment pv_2 intersects segment v_1v_3, see Figure 3.18.

---

[1] Three points on the sphere are called collinear if there exists a great circle on which they all lie. The centroid of a set of points $p_1, p_2, \ldots p_N$ points is their arithmetic mean $(p_1 + p_2 + \ldots + p_N)/N$.

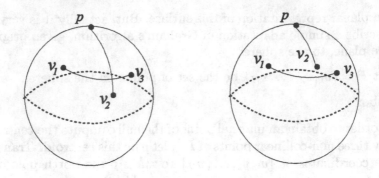

Figure 3.18.   Testing whether v_2 is a hull vertex.

- if $pv_2 \cap v_1v_3 \neq \emptyset$ continue the test with the triple $\{v_2, v_3, v_4\}$.
- if $pv_2 \cap v_1v_3 = \emptyset$ remove v_2 and test the triple $\{v_0, v_1, v_3\}$.

The scan terminates when it advances all the way around to reach the starting point.

In a similar way one can adapt other planar convex hull algorithms for computing the m-convex hull on the sphere of a set of points in Euclidean position. So, as in the cases of cylinder and torus, we can describe roughly the algorithm on the sphere as shown in Table 3.3.

ALGORITHM CH-SPHERE(P)

INPUT: $P = \{v_1, v_2, \ldots, v_N\}$ *set of points on the sphere.*

OUTPUT: *Metrically convex hull of P in the sphere.*

STEP 1 *If P is in Euclidean position go to Step 3.*

STEP 2 *The m-convex hull is the whole sphere. (END)*

STEP 3 *Compute the convex hull of P in the sphere, adapting a planar convex hull algorithm.(END)*

Table 3.3.   Algorithm for computing the metrically convex hull of a set on the cylinder.

THEOREM 3.8 *Algorithm* CH-SPHERE(P) *computes the m-convex hull of a set P of N points on the sphere in optimal $O(N \log h)$ time, h being the number of points in the boundary of the convex hull.*

Proof: This is straightforward since we have that Step 1 requires linear time, just deciding the separability of the set with the center of the sphere, using, for instance, Megiddo's algorithm (see [Megiddo, 1983]); and Step 2 can be done in constant time. Regarding Step 3, it suffices to adapt an optimal planar convex hull algorithm to carry out this step in $O(N \log h)$ time ([Seidel, 1997, Kirkpatrick and Seidel, 1986, Chan et al., 1995, Chan, 1995]). □

3.4 METRICALLY CONVEX HULL ON THE CONE

In order to compute the m-convex hull of a set of N points on the cone we will use the representation of it as the metric space $X = \mathcal{C}_1 \cup \mathcal{C}_2 \cup \mathcal{C}_3$ built by identifying three copies of its developed region, see the Chapter 'Preliminaries' and Figure 3.19.

As in the case of the surfaces mentioned above, when the set of points is in Euclidean position the behavior of this surface will be similar to that in the plane; otherwise we have proved:

LEMMA 3.5 *The m-convex hull of a set of three points in non-Euclidean position on the cone must contain the vertex of the cone.*

Proof: Let $P = \{v_1, v_2, v_3\}$ be a set in non-Euclidean position on the cone; let Γ be the curve defined by the segments joining these points pairwise, and v the vertex of the cone. *A priori*, we know that P and Γ are both contained in the hull of P, $C_m(P)$. Define $r = \min\{d(p, v);\ p \in C_m(P)\}$. Suppose that $r > 0$ and draw the circumference centered at v with radius r. In the same way as in Lemma 3.2 we can prove that the shaded region in Figure 3.20 must be contained in the hull, but this contradicts that $r = \min\{d(p, v);\ p \in C_m(P)\}$, so it must be $r = 0$ and $v \in C_m(P)$.

□

As in the case of the sphere, for computing the m-convex hull of $P = \{v_1, v_2, \ldots, v_N\}$ on the cone, if P is in non-Euclidean position, we adapt a planar convex hull algorithm for computing the convex hull of $\phi^{-1}(P)$ in X (remember that X might be non-embeddable in the plane, but is locally isometric to it). In order to adapt Graham's scan, for instance, it suffices to sort the points of $\phi^{-1}(P)$ lexicographically by polar angle and distance from the cone's vertex, then to start the scan using the same criterium used in the sphere, see Section 3.3. We will

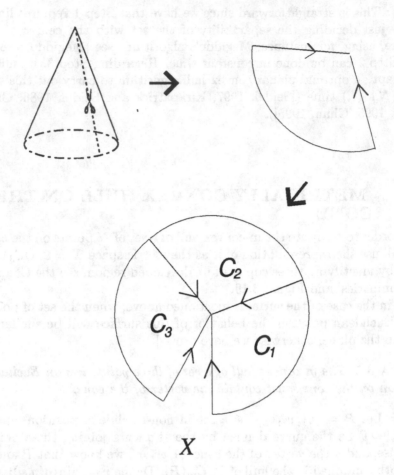

Figure 3.19. The metric space X is built from three copies of the developed region of the cone

obtain a convex region in the Euclidean sense, let us call that region $CH(\phi^{-1}(P))$; that is, the segments joining two points on $CH(\phi^{-1}(P))$ is contained in it. Just considering the region $CH(\phi^{-1}(P)) \cap C_2$, one has an m-convex region in the cone containing P. It suffices to remove from the border of $CH(\phi^{-1}(P)) \cap C_2$ those arcs that are not minimum geodesics joining two points, and we have the m-convex hull of P in the cone, see Figure 3.21.

So we can describe roughly the algorithm on the cone in the way Table 3.4 shows.

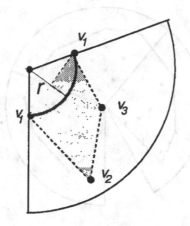

Figure 3.20.   The shaded region must be contained in $C_m(P)$

ALGORITHM CH-CONE(P)

> INPUT: $P = \{v_1, v_2, \ldots, v_N\}$ *set of points on the cone.*
>
> OUTPUT: *Metrically convex hull of P on the cone.*
>
> > STEP 1 *If P is in Euclidean position go to Step 3.*
> >
> > STEP 2 *Compute the convex hull of $\phi^{-1}(P)$ in X adapting a planar convex hull algorithm. The m-convex hull of P is $CH(\phi^{-1}(P)) \cap C_2$. (END)*
> >
> > STEP 3 *Compute the convex hull of P in the cone using a planar convex hull algorithm. (END)*

Table 3.4.   Algorithm for computing the metrically convex hull of a set on the cone.

The analysis of this algorithm is similar to the algorithms on surfaces above, so we can claim:

THEOREM 3.9 *Algorithm* CH-CONE(P) *computes the m-convex hull of a set P of N points in the cone in optimal $O(N \log h)$ time, h being the number of points in the boundary of the convex hull.*

Finally, we must remark that the result presented in Lemma 3.5 holds for hyperconvexity:

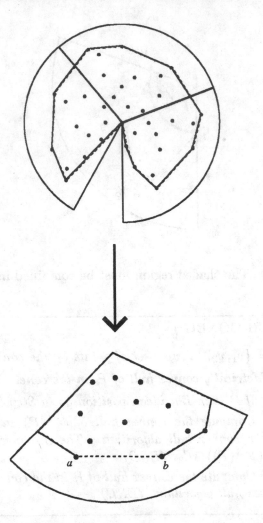

Figure 3.21.   The region in the central copy is the m-convex hull on the cone. Note that we must remove the line joining a and b because this line is not a segment.

LEMMA 3.6 *The hyperconvex hull of a set of two points on the cone must contain the vertex of the cone.*

The proof of this result follows easily by combining the proofs of Lemmas 3.1 and 3.5.

So we can obtain the hyperconvex hull of a set on the cone in the same way as we have obtained the m-convex hull, but removing no lines on the border of the region in the central copy. We will have that:

THEOREM 3.10 *Algorithm* CH-CONE(P) *can be adapted for computing the hyperconvex hull of a set P of N points on the cone in optimal $O(N \log h)$ time, h being the number of points in the boundary of the convex hull.*

4.   ANALYSIS OF COMPLEXITY

Although the complexity of our algorithms in the worst case, as we have seen, is the same as in the plane, we will analyze the complexity of them in the average case. We will show here that our algorithms run almost always in linear time.

Recall that in the first step of the CH-CYLINDER(P) algorithm we ask whether P are contained between two opposite generatrices; when the answer is YES we must use a convex hull algorithm in the plane, and, as is well known the average case performance analysis in this case (see, for instance, [Preparata and Shamos, 1985]); but when the answer is NO computing the m-convex hull is equivalent to computing the extremes of a set of N points, and this is linear. Moreover, as shown in Proposition 2.1, in most of the situations the points will be in Euclidean position. In fact:

PROPOSITION 3.1 *If N points are chosen uniformly and independently at random from a cylinder the expected time used by* CH-CYLINDER(P) *is $O(N) + (1 - N/2^{N-1})O(N) + N/2^{N-1}O(N \log \log N)$.*

Proof: Since one can decide in linear time whether P is in Euclidean position or not, the result follows from Proposition 2.1 and from [Rényi and Sulanke, 1963] (See [Preparata and Shamos, 1985]).   □

Proposition 3.1 says that in most cases it is possible to compute the m-convex hull in linear time, but in most of those cases the convex hull so obtained is too big; in fact, with the probability given in Proposition 2.1 the hyperconvex hull and the m-convex hull will agree.

In the cases of the sphere or the torus we have results similar to that expressed in Proposition 3.1, but using now Proposition 2.4 or 2.3 instead of Proposition 2.1.

PROPOSITION 3.2 *If N points are chosen uniformly and independently at random from a torus, the expected time used by* CH-TORUS(P) *is*

$$O(N) + \frac{N}{2^{2N-2}}(2^N - 2N)O(N) + \frac{N^2}{2^{2N-2}}O(N \log \log N) + (1 - \frac{N}{2^{N-2}}(1 - \frac{N}{2^N}))O(1).$$

PROPOSITION 3.3 *If N points are chosen uniformly and independently at random from a sphere, the expected time used by* CH-SPHERE(P) *is*

$$O(N) + \frac{N^2 - N + 2}{2^N}O(N \log \log N) + (1 - \frac{N^2 - N + 2}{2^N})O(1).$$

Thus, as in the cylinder, the probability of computing the m-convex hull in the sphere or the torus in linear time is extremely high.

5.   MINIMUM ENCLOSING POLYGON

As we have seen in previous sections, the m-convex hull of a point set on the cylinder does not share another planar property and it is not, in general, the minimum enclosing polygon of the point set (we define the *minimum enclosing polygon* of a point set as the shortest polygon enclosing the set). In fact, observe that the minimum enclosing polygon of a set that is not in Euclidean position is not metrically convex (see Figure 3.22).

However, it is possible to find that polygon in optimal time on the cylinder using a planar dynamic convex hull algorithm [Hershberger and Suri, 1996]. In that work it is shown that a sequence of n operations, each an insertion, or deletion, can be processed in time $O(n \log n)$, provided all the operations are known in advance. So we develop the cylinder by a generatrix and compute the convex hull of the planar point set so obtained. In each new step we now delete the leftmost point and insert it on the right. In other words, that generatrix sweeps the cylinder from left to right. In each one of those steps we compare the length of the convex hull so obtained in order to obtain the minimum. Thus we can describe the algorithm for the minimum enclosing polygon in the way shown in Table 3.5.

THEOREM 3.11 *Algorithm* MIN-POL(P) *computes the minimum enclosing polygon of a point set on the cylinder P in optimal $O(n \log n)$ time.*

Proof: It is clear that there exists a generatrix on the cylinder that does not intersect the minimum enclosing polygon of a point set. Developing the cylinder by that generatrix we obtain a planar set such that the

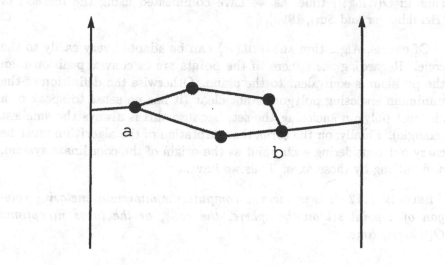

Figure 3.22. The minimum enclosing polygon of a point set is not always m-convex (the segment between a and b is not included in that polygon).

ALGORITHM MIN-POL(P)

   INPUT: $P = \{v_1, v_2, \ldots, v_N\}$ set of points on the cylinder.

   OUTPUT: *Minimum enclosing polygon of P on the cylinder.*

     STEP 1 *If P is in Euclidean position go to Step 5.*

     STEP 2 *Sort the points of P circularly by their abscissae, and call the list starting at the i-th element C_i.*

     STEP 3 *Compute the convex hull of each C_i dynamically in the plane and compare its length with the minimum obtained in previous steps.*

     STEP 4 *Report the minimum obtained in Step 3. (END)*

     STEP 5 *Compute the convex hull of P on the cylinder using a planar convex hull algorithm. (END)*

Table 3.5. Algorithm for computing the minimum enclosing polygon.

minimum enclosing polygon is its convex hull. Finally, the algorithm runs in $O(n \log n)$ time, as we have commented using the method of [Hershberger and Suri, 1996]. □

Of course, Algorithm MIN-POL(P) can be adapted very easily to the cone. Regarding the sphere, if the points are in convex position then the problem is equivalent to the plane. Otherwise the definition of the minimum enclosing polygon is not clear (it has no sense to speak of a shortest polygon enclosing the set, because this is always the smallest triangle). Finally, on the torus the adaptation of the algorithm must be carry out considering each point as the origin of the coordinate system, and cutting by those axes. Thus we have:

THEOREM 3.12 *It is possible to compute the minimum enclosing polygon of a point set on the sphere, the cone, or the torus in optimal $O(n \log n)$ time.*

6.   NOTES AND COMMENTS

We have that, in any case, it is possible to compute the m-convex hull of a set of N points on the cylinder and on the torus in optimal time $O(N \log h)$; moreover, in most cases our algorithms run in linear time. But when the points are in non-Euclidean position on the cylinder, or in non-cylindrical position on the torus, the m-convex hull is too big for many purposes. It is then interesting to find other structures in order to use them to solve problems such as the diameter of a set in those surface, for instance. This problem has been solved in the plane, in optimal time $O(N \log N)$ using the convex hull as a previous step (the diameter is always obtained in extreme points), but one cannot do this in the cylinder as is shown in Figure 3.23.

Figure 3.23 shows a set of points on the cylinder such that the diameter of the set is obtained by points that are not on the border of the m-convex hull of the set.

In general we will see throughout this book that those structures exist, but that the role played in the plane by the convex hull as an ubiquitous tool which appears almost everywhere is not played by any other structure; so we will have to use different tools according to each particular problem.

Summarizing, we think that the convex hull is not a useful tool for working in surfaces, because, in general, this set will be too big. We are convinced that this fact is true for almost all kinds of surfaces except for

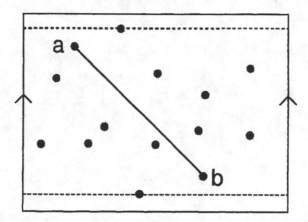

Figure 3.23. The diameter of this set is the distance between *a* and *b*, but these points are not on the boundary of the convex hull of the set.

those which are very flat. However, for collections of points that are not very far apart from each other the convex hull will preserve the shape that it presents in the plane.

Some of the material in this chapter has appeared previously in [Dana et al., 1997], but we have included here some unpublished results such as those regarding the cone and those contained in Sections 4 and 5.

Chapter 4

VORONOI DIAGRAMS

The Voronoi diagram is one of the most versatile structures in Computational Geometry. It has been considered second in importance only to the convex hull. Without doubt the reason for this assessment is that Voronoi diagrams have applications and are used extensively in a great variety of disciplines (see [Aurenhammer, 1991, Okabe et al., 1992]). So it is possible to say that the Voronoi diagram is an interdisciplinary concept, and, in fact, it has independent roots in many fields.

Basically, this structure has been studied to model proximity relations between objects (the closest, the furthest, ...). Therefore we think that it is important to introduce Voronoi diagrams in surfaces. Anyway, unlike other problems in Computational Geometry, closest point Voronoi diagrams have been considered previously by other authors in some depth. So in this chapter we will summarize their results about closest point Voronoi diagrams and furthest point Voronoi diagrams and will introduce a generalize Voronoi diagram that will be important later as a tool in triangulations.

1. INTRODUCTION

Although some authors cite Descarte's works in cosmic fragmentation *Le Monde de Descartes, ou Le Traité de la Lumiére* and *Principia Philosophiae* (both published in 1644) as the first works on Voronoi diagrams, it seems to be clear that the first undisputed studies of the concept appear in the work of Dirichlet [Dirichlet, 1850] and Voronoi [Voronoi, 1908]. In fact two of the many labels used to described the diagram are *Voronoi diagram* and *Dirichlet tessellation*. These two works were in a

61

mathematical context, but since then Voronoi diagrams have been considered (and rediscovered) in many other areas. So it is possible to find that structure not only in mathematical texts, but also in books or articles, on Crystallography [Niggli, 1927, Nowacki, 1976, Schaudt and Drysdale, 1991], Meteorology [Thiessen, 1911, Horton, 1917, Whitney, 1929], Geology [Popoff, 1966, Harding, 1921, Harding, 1923], Physics [Wigner and Seitz, 1933], Agricultural Science [Brown, 1965], Biology or Ecology [Bartllet, 1975, Pielou, 1977, Getis and Boots, 1978], Geography, Astronomy [V. Icke and R. Van de Weygaert, 1987], Urban and Regional Planning [Shieh, 1985, Bogue, 1949, Dacey, 1965, Snyder, 1962], etc. (we refer the reader to the book [Okabe et al., 1992] for a good historical introduction on the subject).

If we restrict ourselves to the field of Computational Geometry the Voronoi diagram (or its dual, the Delaunay triangulation) has been used as a preprocessing for other problems as the nearest neighbors, all nearest neighbors, minimum spanning tree, the traveling salesperson problem, etc. We will see in this chapter how the Voronoi diagram can be used as a preprocessing for those problems on our surfaces.

The structure of this chapter is as follows. We will start in Section 2 summarizing the known results on the closest point Voronoi diagram of a set of sites on our surfaces, i.e., the structure that assigns to each site the locus of points at least as close to that site as to any other site. Mainly the techniques used to calculate closest-point Voronoi Diagrams of a set of sites on a surface are based on computing the Voronoi diagram of an associated set of points in the plane, but we will see in Section 4 that this technique cannot be applied to calculate the furthest point Voronoi Diagram, thus this diagram must be computed directly using either a divide and conquer approach or an incremental one. In any case, up to now the complexity of the algorithms is the same as in the plane. The first example of a structure that requires more effort on surfaces than in the plane comes from considering some generalized Voronoi diagrams as the polar diagram (used in some problems of angle optimization) that needs quadratic time in the cylinder and only $O(N \log N)$ time in the plane.

2.   VORONOI DIAGRAMS

Given a set of distinct sites $S = \{P_1, P_2, \ldots, P_N\}$ in a metric space X we call the *Voronoi region* of P_k, $1 \leq k \leq N$, (denoted Vor(P_k)) the locus of points in X at least as close to P_k as to any other site,

$$\text{Vor}(P_k) = \{Q \in X \; ; \; d(Q, P_k) \leq d(Q, P_j), \; j \neq k\}.$$

The set given by

$$\mathcal{V} = \{\text{Vor}(P_1), \ldots, \text{Vor}(P_N)\}$$

is a tessellation of X called the *Voronoi diagram* generated by S (or more commonly, the Voronoi diagram of S), and denoted $\text{Vor}(S)$.

This structure has been studied on the surfaces considered here by several authors, and so Takeda [Takeda, 1985] considered cylindrical Voronoi diagrams. Miles [Miles, 1971], Brown [Brown, 1980], Paschinger [Paschinger, 1982] and Ash and Bolker [Ash and Bolker, 1985] have studied spherical Voronoi diagrams. A Voronoi diagram on a torus is discussed by Upton and Finglenton [Upton and Fingleton, 1985], and even the conic Voronoi diagram has been studied, mainly by R. Klein [Dehne and Klein, 1987, Klein, 1988, Klein and Wood, 1988].

All these structures can be considered as a particular case of the Voronoi diagram on orbifolds introduced by Mazón and Recio [Mazón, 1992, Mazón and Recio, 1997]. In those works they study Voronoi diagrams for Euclidean and spherical two–orbifolds. This class of surfaces includes all the locally Euclidean and locally spherical surfaces such as the cylinder, the torus, the Möbius band, the projective plane, the Klein bottle, the cone, etc. Basically they develop those surfaces in the plane (or the sphere), and compute Voronoi diagrams on those spaces by considering enough copies of that planar representation of each space (three in the case of the cylinder or the cone, nine in the case of the torus or the Klein bottle — see Chapter 1 —). Thus the initial set of N sites S is transformed in a set of kN sites in the plane (where $k = 3$ in the case of the cylinder and the cone, or $k = 9$ in the case of the torus or the Klein bottle); now the Voronoi diagram of those kN sites is computed and the intersection of this diagram with the central copy of the planar representation of the original space gives the Voronoi diagram on the surface just by removing the edges between regions of equivalent points. This procedure leads to an optimal $O(N \log N)$ algorithm (see Figure 4.1).

Although it is possible to find the description of all those diagrams in [Okabe et al., 1992], we will summarize now some aspects of them that will be used in other chapters of this book.

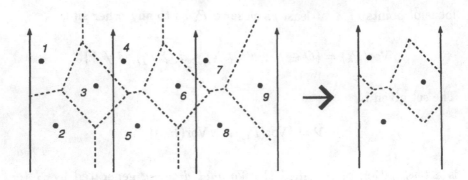

Figure 4.1. How to get the cylindrical Voronoi diagram from a planar Voronoi diagram.

2.1   VORONOI DIAGRAMS ON THE CYLINDER

Using Mazón and Recio results [Mazón, 1992, Mazón and Recio, 1997], we know how to compute Voronoi diagrams in the cylinder. But sometimes it is useful to know how to compute bisectors directly and some of their properties. Thus, it is easy to show the following result:

LEMMA 4.1 *Given two points P and Q on the cylinder their bisector is the union of the two geodesics with least length joining a point in the opposite generatrix from P and the other in the opposite generatrix from Q.*

We will see now that the converse of this result is also true:

LEMMA 4.2 *Let M_1 and M_2 be two points on the cylinder not on the same generatrix. Then there exist two points P and Q such that the union of the two geodesics with least length joining M_1 and M_2 is the bisector of P and Q.*

Proof: If h_1 and h_2 are the two geodesics joining M_1 and M_2 (see Figure 4.2). Then P and Q lie on the generatrices g_1 and g_2 opposite to M_1 and M_2 respectively.

In order to fix the position of Q in g_2 note that h_1 and h_2 with a half of the generatrix starting in M_2 is the planar Voronoi diagram of the points P, Q and the copy of Q to the right of P as Figure 4.3 shows.

If we now call γ the angle defined by h_1 and the line joining Q with M_2, and call ϕ the angle marked in Figure 4.4, as is proved in [Ash and Bolker, 1985] $\gamma + \phi = \pi$.

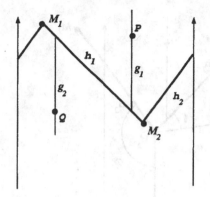

Figure 4.2. The points P and Q lie on the generatrices g_1 and g_2 opposite to M_1 and M_2.

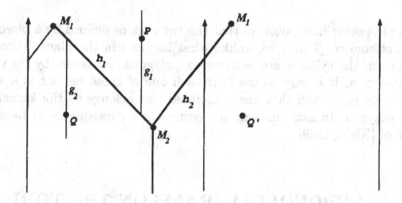

Figure 4.3. h_1, h_2 and a half of the generatrix starting in M_2 is the planar Voronoi diagram of the points P, Q and the copy of Q.

As ϕ is known it is enough to consider the line making an angle γ with h_1 from M_2 and the point of intersection of this line with g_2 will be Q; the computation of the position of P is now straightforward.

□

In the previous lemma we can notice a similarity with the planar case, since bisectors in the plane (straight lines) are defined by two points, and the same happens on the cylinder; however, in the plane any pair of points in the bisector characterizes it and on the cylinder only a very specific pair characterizes a bisector.

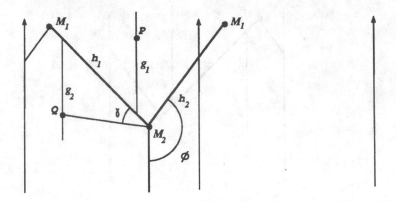

Figure 4.4.   The angles γ and ϕ.

On the other hand, observe that the two regions defined by a bisector are not convex. Therefore, unlike what happens in the plane, Voronoi regions in the cylinder are not convex polygons. However, by its very construction, it is easy to see that each one of those regions is a *star shaped* polygon, and that the associated site belongs to the kernel of that polygon. In fact, this can be deduced as a consequence of Lemma 1.2.6 of [Klein, 1989].

2.2   VORONOI DIAGRAMS ON THE TORUS

As we have said above, in order to calculate the Voronoi diagram of a set of N sites on the torus we develop the torus in the plane and consider nine copies of it (surrounding the central copy). Thus the set of N sites on the torus is transformed into a set of $9N$ sites in the plane, computing the Voronoi diagram of this planar set, the intersection of this diagram with the central copy of the planar representation of the torus gives the Voronoi diagram of the original set on the torus just by removing the edges between regions of equivalent points. This procedure leads to an optimal $O(N \log N)$ algorithm (see Figure 4.5).

Obviously, bisectors on the torus are more complicated than on the cylinder. In fact, they are not connected but they can be obtained by making two copies of a cylindrical bisector (see Figure 4.6).

As in the cylinder, a bisector does not divide the torus into two convex regions. And again all Voronoi regions are star shaped. But it is

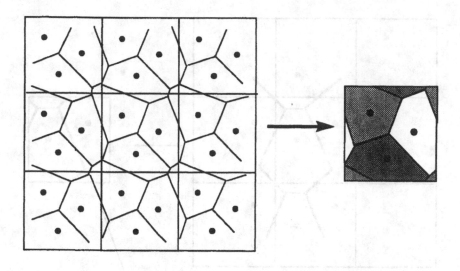

Figure 4.5.   How to get a toroidal Voronoi diagram.

easy to see from the construction above that if the points are neither in Euclidean position nor in cylindrical position and they are uniformly situated on the torus, then all Voronoi regions are convex.

2.3   VORONOI DIAGRAMS ON THE SPHERE

The spherical Voronoi diagram is potentially useful to consider global problems such as the location of airline hubs or the optimization of market places. Observe now that the bisector between two points is a great circle (which passes perpendicularly through the mid-point of the segment joining those two points) that divides the sphere into two disjoint hemispheres. Therefore Voronoi regions are always convex polygons.

Practically all planar algorithms can be adapted for computing spherical Voronoi diagrams, but one of the easiest methods for computing that structure is the one proposed by Brown [Brown, 1980]. He considers the N sites on the sphere as points in Euclidean 3-dimensional space and computes the convex hull (in optimal $O(N \log N)$ time) of that finite collection of sites. Now, it is not difficult to see that the edges and faces of this convex hull give the dual of the Voronoi diagram on the sphere. Finally, given the dual, the Voronoi diagram can be produced in only $O(N)$ additional time. The validity of this method is based on the

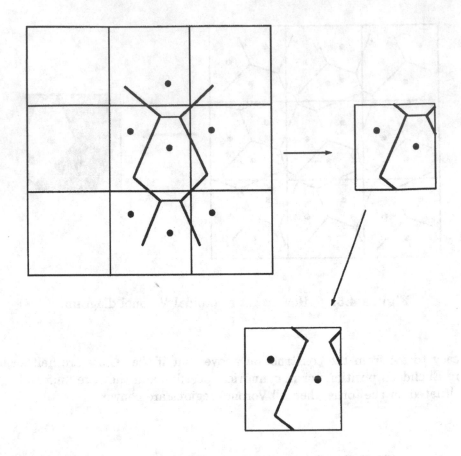

Figure 4.6. A toroidal bisector can be obtained from nine copies of the planar representation of the torus.

property of the vertices of the spherical Voronoi diagram which are the centers of the circles which pass through three of the sites and do not contain any other site. And it can be checked that those circles determine the planes that bound the faces of the (Euclidean 3-dimensional) convex hull of the N sites.

2.4   VORONOI DIAGRAMS ON THE CONE

Dehne and Klein [Dehne and Klein, 1987] generalize the planar sweep-circle technique of the plane to work on a cone. Moreover, Mazón and

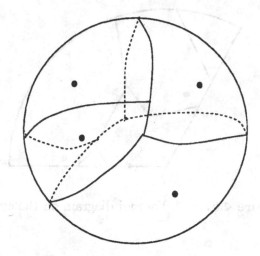

Figure 4.7.   A Voronoi diagram on the sphere.

Recio [Mazón, 1992, Mazón and Recio, 1997] point out that their procedure (described at the beginning of this section) can be easily adapted to computing Voronoi diagrams on the surface of a cone of arbitrarily angle α, whenever 3α does not exceed 2π. In fact, they introduce that restriction because they want to use a planar method, but almost all planar methods for computing Voronoi diagrams can be adapted very easily to the surface X obtained by gluing three copies of the cone defined in Chapter 1, so that extra condition can be removed easily and their technique can be used in any case, even when 3α does exceed 2π (it is enough to compute the Voronoi diagram in X and then to choose only the central copy of the cone).

3.   PROXIMITY PROBLEMS AND VORONOI DIAGRAMS

It is a classical result that Voronoi diagrams provide the information needed to compute in optimal time many proximity problems (see, for instance, [Preparata and Shamos, 1985]). Amongst them we have the *closest pair* (Given N sites, report two whose mutual distance is smallest), *all nearest neighbors* (now, find a nearest neighbor of each),

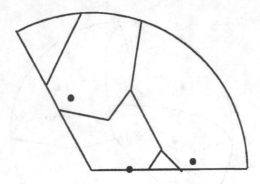

Figure 4.8.   A Voronoi diagram on the cone.

the *minimum spanning tree* (construct a tree of minimum length whose vertices are the given sites).

In fact, to solve these problems we need only examine the edges of the dual of the Voronoi diagram, which is called the Delaunay triangulation in the plane. In order to consider the dual of the Voronoi diagram on our surfaces we join adjacent points p and q in the Voronoi diagram by a segment on the surface.

Unfortunately, in general the dual of a Voronoi diagram on our surfaces can contain two segments that cross each other, and therefore is not a triangulation (see Figure 4.9 — we will come back to this point in Chapter 7—). But we can solve all proximity problems that we mentioned above just by removing those edges.

Thus, given a set of sites S on a surface, we call the *Euclidean dual* of Vor(S) (denoted Ed(S)) to the subgraph of the dual of Vor(S) defined by removing all the edges (of the dual) with extremes p and q that cannot be deformed homotopically into Jordan curves joining those points in Vor(p) $\cup$ Vor(q).

For the sake of simplicity we will call S one of our surfaces, i.e., the cylinder, the sphere, the torus, or the cone. Figure 4.10 shows the Euclidean dual of the Voronoi diagram depicted in Figure 4.9. Observe that the bold segment in Figure 4.9 cannot be deformed into a Jordan curve joining the same extremes within their Voroni regions.

The next result proves that for many proximity problems the Euclidean dual plays the role on surfaces of the Delaunay triangulation in the plane.

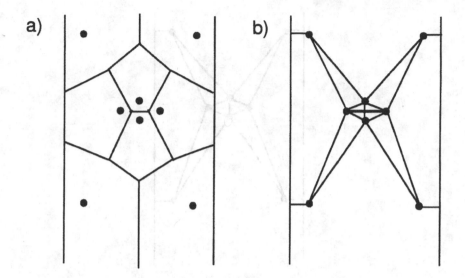

Figure 4.9. The dual (b) of this Voronoi diagram (a) is not a triangulation

LEMMA 4.3 *Let S be a set of sites in S. For any partition $\{S_1, S_2\}$ of S the shortest segment between points of S_1 and points of S_2 belongs to $Ed(S)$.*

Proof: If the shortest segment between a point a of S_1 and a point b of S_2 does not belong to $Ed(S)$. Then it must be crossed by another segment with extremities a' and b'. We then have a quadrilateral $aa'bb'$. But in that case the diagonal ab is larger than the diagonal $a'b'$ and than the edges of the quadrilateral; therefore ab cannot be the shortest segment between points of S_1 and points of S_2. □

In other words, this lemma tells us that we need only examine the edges of $Ed(S)$ to solve the proximity problems mentioned.

COROLLARY 4.1 *Given a set of sites S the closest pair, all nearest neighbors, and the minimum spanning tree are subgraphs of $Ed(S)$.*

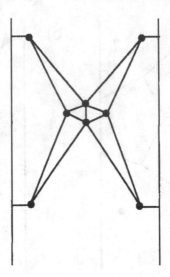

Figure 4.10.   The Euclidean dual of Figure 4.9.

3.1   VORONOI DIAGRAMS AND CONVEX HULLS

Regarding the relationship of the Voronoi diagram to the convex hull, note that, in the cylinder for a set that is in non-Euclidean position, a Voronoi region is unbounded if and only if its corresponding site is a point on the boundary of the m-convex hull. So the m-convex hull on the cylinder can be obtained in linear time from the Voronoi diagram. This relationship holds for the cone as well (the other unbounded surface), but it has no meaning in the cases of the sphere or the torus.

Finally, as we can see in Figure 4.10, the Euclidean dual is not a triangulation. So probably one of the most interesting consequences of Voronoi diagrams in the plane can not be solved yet. We will study triangulations in the last chapter.

4.   FURTHEST POINT VORONOI DIAGRAM

Given a set of distinct sites $S = \{P_1, P_2, \ldots, P_N\}$ in a metric space X we call *the furthest point Voronoi region of* P_k, $1 \leq k \leq N$ (denoted $\text{Vor}^F(P_k)$) the locus of points in X at least as far to P_k as to any other site,

$$\mathrm{Vor}(P_k) = \{Q \in X \; ; \; d(Q, P_k) \geq d(Q, P_j), \; j \neq k\}.$$

The set given by

$$\mathcal{V}^F = \{\mathrm{Vor}(P_1), \ldots, \mathrm{Vor}(P_N)\}$$

is a tessellation of X called the furthest point Voronoi diagram generated by S (or more commonly, the furthest point Voronoi diagram of S), and is denoted $\mathrm{Vor}^F(S)$.

In [Brown, 1980] it is observed that computing the furthest point Voronoi diagram of a set of sites on the sphere is equivalent to computing the closest point Voronoi diagram of the set defined by the antipodal points of the original sites. In fact, this argument can be considered from a more general point of view, as we now see.

Given a metric space X we say that y is the *antipodal* of x if y is the single farthest point to x. Therefore, if in X all points have antipodals, we can define a map ant : $X \to X$ such that ant maps each point to its antipodal, and so next theorem is a straightforward generalization of Brown's argument.

THEOREM 4.1 *Let S a finite set of sites in a metric space X, such that the map ant : $X \to X$ is an isometry, then*

$$Vor^F(S) = Vor(ant(S)).$$

Proof: It is easy to check that under the hypotheses of the theorem, if we apply the map *ant* to a triple of points the relation of proximity between those three points is reversed, using this observation the proof is immediate □

Under its hypotheses Theorem 4.1 provides a method for computing the furthest point Voronoi diagram not only on the sphere but on the torus and, in general, on any other surface. Obviously the last condition (that map ant is an isometry) is not superfluous, as Figure 4.11 shows.

Clearly, in the cylinder the map ant is not defined (when it is considered as an unbounded space) or is not an isometry (when it is considered as a compact space) and in this last case the same happens as in Figure 4.11; therefore we cannot use Theorem 4.1 for computing the furthest point Voronoi diagram in the cylinder.

Another possible attempt could be to mimic the method used in [Mazón, 1992, Mazón and Recio, 1997] for computing the closest point Voronoi diagram (see Section 2). In that method, given N sites on a cylinder, we develop the cylinder in the plane and consider three copies of it (one to

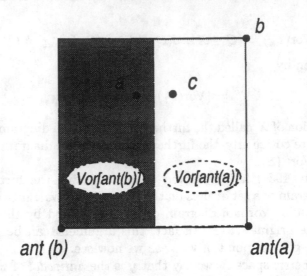

Figure 4.11.   The map ant must be an isometry: c is in $\text{Vor}(\text{ant}(a))$ but it is closer to a than to b.

each side of a the central copy). Thus the set of N sites on the cylinder is transformed into a set of $3N$ sites in the plane, computing the closest point Voronoi diagram of this planar set, the intersection of this diagram with the central copy of the planar representation of the cylinder gives the Voronoi diagram of the original set just by removing the edges between regions of equivalent points.

But again this method cannot be used now for computing furthest point Voronoi diagram, since we can find situations such as that described in Figure 4.12.

Therefore we are going to use one of the classical design paradigms used in Computational Geometry, the divide and conquer, for computing the cylindrical furthest point Voronoi diagram. Hence, as happens in the plane, the key to this method is to construct the dividing chain, that will be used in the merge process. Moreover, we have to guarantee that that dividing chain can be constructed in linear time in order to obtain an optimal algorithm. In what follows we describe the whole method.

1. Split the original set of sites S into two subsets S_1 and S_2 of approximately equal sizes by a parallel $p : \{y = K\}$; we will assume that S_1 is above the parallel and that S_2 is under the parallel.

2. Construct the furthest point Voronoi diagram of S_1 and S_2 recursively. Each one of those two diagrams defines segments in the par-

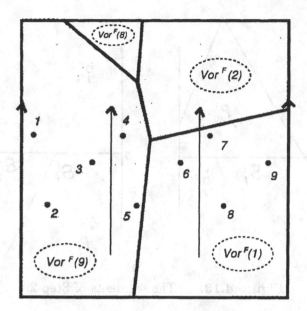

Figure 4.12.   In this example site 6 would be in $\mathrm{Vor}^F(1)$, but actually, site 7 is a copy of 1, and therefore sites 6 and 1 are very close.

allel p (the intersection of the regions with the parallel). That is to say, if $S_1 = \{P_1^1, P_2^1, \ldots, P_{n_1}^1\}$ and $S_2 = \{P_1^2, P_2^2, \ldots, P_{n_2}^2\}$, we call

$$S_j^i = \{(x, K) \; ; \; (x, K) \in Vor^F(P_j^i)\}$$

with $i = 1, 2$ y $j = 1, 2, \ldots, n_i$.

3. *(Constructing the dividing chain).*

(a) Start from a point (x_0, K) in p. For this point there exist values r and s such that $(x_0, K) \in S_r^1 \cap S_s^2$. Compute the bisector between P_r^1 and P_s^2.

(b) If this bisector has no intersection with p in a point of $S_r^1 \cap S_s^2$, we move to the right along p until $S_{r'}^1$ or $S_{s'}^2$ is reached, and we ask again whether the new bisector so constructed has an intersection with p inside the intersection of the corresponding regions.

(c) If we again reach the initial point in p, that would mean that we have not reached the dividing chain, i.e., the parallel p does not intersect the dividing chain, or, in other words, that chain is completely contained in one of the half cylinders defined by p. In this latter case we proceed as follows

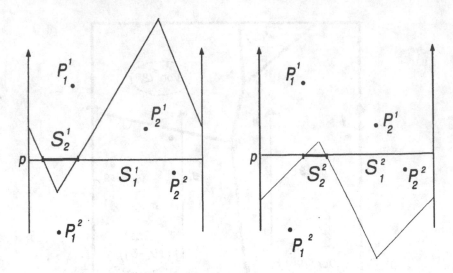

Figure 4.13. The segments of Step 2.

i. It is possible to decide in constant time the half cylinder in which the dividing chain lies. For that purpose it is enough to compare the distances of the initial point (x_0, K) with the points P_r^1 and to P_s^2 given in Step 3a. And the chain will be above the parallel p if the distance to P_r^1 is bigger than the distance to P_s^2, and under the parallel otherwise.

ii. Once we know the half cylinder in which the dividing chain lies, we find that chain by walking along the generatrix of the initial point $(x = x_0)$. In order to find that polygonal line, we apply a test similar to that of Step 3a, hence we seek a point of the generatrix $x = x_0$ verifying that if that point is in $\text{Vor}^F(P_r^1) \cap \text{Vor}^F(P_s^2)$ then the bisector between P_r^1 and P_s^2 intersects $x = x_0$ inside $\text{Vor}^F(P_r^1) \cap \text{Vor}^F(P_s^2)$. (We know that we will reach the dividing chain because that polygonal line is non-trivial from a homotopical point of view, it wraps around the cylinder).

(d) Once we have a point on a bisector that is part of the dividing chain (as in the plane) we move along that bisector until it crosses an edge of one of the polygons which the bisector splits, and then we change one of the original points and we follow along the new bisector until we again change the direction. This procedure will continue until we encounter the initial point.

Therefore:

LEMMA 4.4 *It is possible to compute the dividing chain between S_1 and S_2, subsets of S linearly separated by the parallel p, in linear time.*

4. Once we have the dividing chain we must discard all edges of $\text{Vor}^F(S_1)$ which are under the chain and all edges of $\text{Vor}^F(S_2)$ that lie above that polygonal line.

Then we conclude:

THEOREM 4.2 *The furthest point Voronoi diagram of N points in the cylinder can be constructed in $O(N \log N)$ time, and this is optimal.*

Of course, the construction of the dividing chain is valid for the closest point Voronoi diagram as well, and thus:

THEOREM 4.3 *The closest point Voronoi diagram of N points in the cylinder can be constructed by divide and conquer in $O(N \log N)$ time, and this is optimal.*

5. GENERALIZED VORONOI DIAGRAMS

In order to facilitate many practical applications in various fields, since the early 1970s the concept of Voronoi diagram has been generalized in many directions. It is not the aim of this book to study each one of those structures on the surfaces that we are considering here. We think that if the reader needs, for a concrete problem, to study a generalized Voronoi diagram on some of those surfaces, taking into account the information provided in this chapter, he can construct by himself that structure. Nevertheless, we present here two of those generalizations, one is the Voronoi diagram of a segment set on the cylinder, and the other is a completely different structure called the polar diagram which in [Grima et al., 1998a, Grima et al., 1998b] has been used as a preprocessing in visibility problems.

In [Okabe et al., 1992] a general framework for those extensions is given in the following terms: Given a space S and a set of N distinct subsets of S $A = \{A_1, A_2, \ldots, A_N\}$, $2 \leq N < \infty$, we consider an assignment of a point P in S to at least one element in A. We put 1 for the pair (P, A_i) if we assign $P \in S$ to A_i, and 0 otherwise. Thus this assignment can be regarded as a mapping from $S \times A$ to $\{1, 0\}$; $\delta : S \times A \to \{1, 0\}$, such that

$$\delta(P, A_i) = \begin{cases} 1, & \text{if } P \text{ is assigned to } A_i, \\ 0, & \text{otherwise.} \end{cases}$$

We call δ an *assignment rule*. And under the assignment rule δ we define

$$V(A_i) = \{P \in S \mid \delta(P, A_i) = 1\}$$

$$e(A_i, A_j) = \{P \in S \mid \delta(P, A_i) = \delta(P, A_j) = 1\} = V(A_i) \cap V(A_j).$$

Every assignment rule δ verifying:

(*i*) Each point P in S is assigned to at least one element of A;

(*ii*) The set $e(A_i, A_j)$ is the boundary of $V(A_i)$, i.e., for any $\varepsilon > 0$ the open ball centered at $P \in e(A_i, A_j)$ and radius ε contains interior points of $V(A_i)$.

forms a tessellation of S called as the *generalized Voronoi diagram* generated by the set A with assignment rule δ in S [Okabe et al., 1992].

5.1   VORONOI DIAGRAMS FOR A SET OF POINTS AND SEGMENTS ON THE CYLINDER

As is well known, an assignment rule can be obtained when we change the generators and consider segments instead of points, equally, we can modify the metric and work, as we have done through out this book, on a surface instead of in the plane. So we obtain a first example of a generalized Voronoi diagram in the Voronoi diagram for a set of points and segments in the cylinder.

We have seen that in order to construct the Voronoi diagram for a set of points on the cylinder it is enough to compute the Voronoi diagram of the set obtained by gluing together three copies of the developed cylinder. But this is not true when the generators are segments on the cylinder, mainly because segments can span over more than one copy of the developed cylinder, as Figure 4.14 shows.

Nevertheless, in [Márquez and Valenzuela, 2000] it is proved that five copies are enough. Combining this fact with the known algorithmic results, it is possible to get:

THEOREM 4.4 *The Voronoi diagram of N points or segments on the cylinder can be constructed in $O(N \log N)$ time, and this is optimal.*

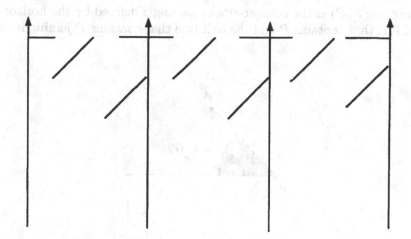

Figure 4.14. Three copies of the cylinder are not enough in the case of Voronoi diagrams for a set of segments.

5.2   POLAR DIAGRAM ON THE CYLINDER

As we have said above, the polar diagram has been used as a pre-processing in visibility problems (more concretely in problems of optimization in visibility such as those considered in [Hurtado, 1993]), or in collision avoidance. Another application of this structure is in some problems related to triangulations of points, as we will see in Chapter 6. Equally, it has been used in the computation of planar convex hulls, obtaining very promising results in the performances of the first implementations of the algorithms.

There are two reasons for introducing this structure in the cylinder, first of all, as we will see, some of its applications can be considered on the cylinder. And, mainly, because it is the first example we can give of a structure that is more time consuming on the cylinder than in the plane.

As we said above, this is a particular example of generalized Voronoi diagram. In the plane the *polar diagram* is defined in the following way (see [Grima et al., 1998a]). Let $A = \{A_1, A_2, \ldots, A_N\}$ be a set of sites in the plane. We consider the following assignment rule, given $P \in \mathbf{E}^2$

$$\delta(P, A_i) = \begin{cases} 1, & \text{if } \text{ang}_{A_i}(P) \leq \text{ang}_{A_j}(P) \text{ for } j \neq i \\ 0, & \text{otherwise} \end{cases}$$

where $\text{ang}_{A_i}(P)$ is the counter-clockwise angle defined by the horizontal half-line that contains P and the half-line that contains P joining it with A_i.

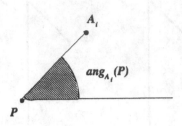

Figure 4.15.   The angle $\text{ang}_{A_i}(P)$.

It is straightforward to see that the assignment rule which we have just described satisfies conditions (*i*) and (*ii*) above; therefore it defines a tessellation of $\mathbf{E}^2$, $V(A, \delta, \mathbf{E}^2)$ called the *Polar Diagram of A*. Intuitively, to say that a point P belongs to $V(A_i)$ means that if we sweep counterclockwise with a half-line having origin at P, the first site we shall find is A_i, i.e., A_i is the site with smallest polar angle with respect to P as origin. In other words, if P is a vehicle equipped with a radar, the first object to appear in its screen will be A_i.

This diagram can be constructed in the plane using either an incremental algorithm or by 'divide and conquer' in optimal time $O(N \log N)$ (check [Grima et al., 1998a]). In the same work that structure is used to compute the convex hull of a set of points in optimal time. We already know that the convex hull is not as practical structure on the cylinder as in the plane, but, nevertheless, we will use the polar diagram in the triangulation problem (given a set of sites in the cylinder, to determine which region in the cylinder can be triangulated using only segments between sites, see Chapter 6).

Therefore we now focus on computing the polar diagram on the cylinder, considering $\text{ang}_{A_i}(P)$ as the counter-clockwise angle defined by the generatrix at P and the half-line which joins P with A_i, see Figure 4.17.

It can be checked that the incremental algorithm can be adapted very easily to compute the polar diagram on the cylinder. But in general the polar diagram for a set of N sites does not have a linear number of vertices, as in the plane, but has $O(N^2)$ vertices, since every pair

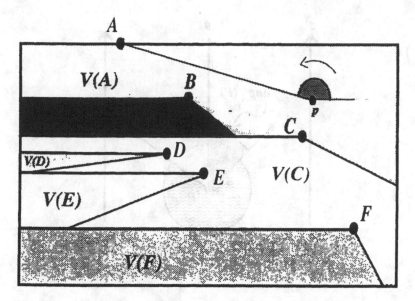

Figure 4.16. A polar diagram in the plane. In this polar diagram we can see that as the point p belongs to the region associate with the site A, then the first site that see p sweeping with a half line counter-clockwise is A.

of points can define a vertex in the polar diagram on the cylinder (see Figure 4.18). Therefore we can obtain the following theorem:

THEOREM 4.5 *It is possible to compute the polar diagram of a set of N sites on the cylinder in optimal $O(N^2)$ time.*

As we pointed out above, this is the first example of a structure that is more complex on the cylinder than in the plane (in the sense that the same structure has more vertices, and, therefore, its worst case analysis does not present the same complexity.

6. NOTES AND COMMENTS

Obviously, a book which has as its main objective the introduction of some of the best known topics of Computational Geometry must contain at least one chapter devoted to Voronoi diagrams. Even more so, these structures are some of the few that have been considered previously by other authors and the complexity of the algorithms are asymptotically the same as the equivalent algorithms for the plane. And, in the contrast

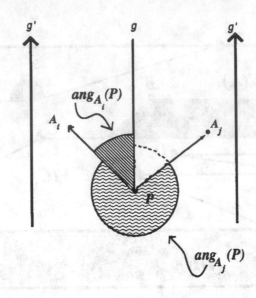

Figure 4.17. If we call g the generatrix containing P, and g' its opposite, P only 'sees to its left and right until it reaches g''.

to what happened with the convex hull, Voronoi diagrams are useful tools (and used) on surfaces, as we will see in the next chapters. For instance, we can use Voronoi diagrams to solve proximity problems, but we have to be careful, because in this respect things are not exactly the same as in the plane. Thus the dual of a Voronoi diagram could not be a triangulation, but it contains a triangulation.

Not only the closest point Voronoi diagram has appeared before in the literature, but the furthest point Voronoi diagram of the sphere was studied in [Brown, 1980]. We have extended the method of computing the furthest point Voronoi diagram not only to the sphere but to any space in which the map *antipodal* is an isometry (as is the case of the flat torus). As the cylinder does not belong to one of those cases we have computed directly by a divide and conquer algorithm its furthest point Voronoi diagram, although some other methods (such as the incremental method) could be valid as well.

Finally, we have presented two examples of generalized Voronoi diagrams, the Voronoi diagram for a set of segments and the polar diagram, and this latter structure presents a complexity on the cylinder higher than in the plane, being this the first example of a structure more complex in the cylinder than in the plane.

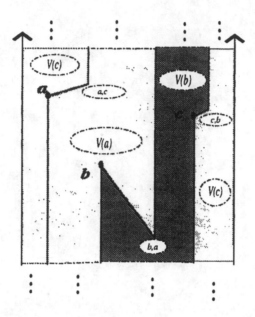

Figure 4.18.   In this polar diagram on the cylinder we can see that each pair of sites has associated a new corner vertex that does not appear in the plane. Those corner vertices occur because a point cannot see a site that is a half cylinder apart.

Obviously there are many ways to follow the line expounded in this chapter. Specific problems will lead to different generalized Voronoi diagrams which have not been considered here. In any case, it is not difficult to follow a parallel construction for higher order Voronoi diagrams on the surfaces considered here.

Chapter 5
RADII

The width, diameter, the circumradius and other similar invariants known as radii are classical functionals which play an important role in convexity theory and in many other applications. In this chapter we study such parameters in the cases of our surfaces. We will see that the usual planar techniques for computing those invariants are not valid in this case and that new methods must be considered.

1.   INTRODUCTION

Classically, in the plane the *width* of a set is the minimum distance between parallel lines of support of the set (a line l is a line of support of a set S if all the set lies in one of the two halfplanes defined by l). The *diameter* is the maximum distance between points of the set, and the *circumradius* is the radius of its minimum enclosing circle.

Although, sometimes the applications of three different parameters associated with sets of sites such as the width, the diameter, and the circumradius are not very similar, those concepts evidently belong to the same family of problems in Computational Geometry (all them are functionals known as radii). To check this last asseveration it is enough to consult [Houle and Toussaint, 1988]. In that paper the authors compute the width of a set of sites in the plane by using two different algorithms, and each of them is an adaptation of a previous algorithm for computing the diameter of sets. In any case, many other papers have treated the three concepts (the width and the diameter obviously, but the circumradius as well) simultaneously, such as in [Chazelle et al., 1993] (see [Goodman and O'Rourke, 1997]).

85

As happens with other tools treated in other chapters, the first reason for treating the width, the diameter, and the circumradius outside Euclidean spaces is that the usual applications of the same problems in the plane (clustering, robotics, Operations Research, etc.—see [Gritzmann and Klee, 1993]) can occur when the points are not in the plane but on other surfaces. But there are two other very important reasons for including a chapter containing, at the least, a common treatment of the width and the diameter in a work like this book. Firstly, the algorithmic solutions of both problems are no longer similar outside the Euclidean case. In same way we could say that the width and the diameter are no longer close relatives (at least from an algorithmic point of view). Secondly, we believe that the solutions which we provide here for solving both problems are very significant as representatives of the methodology that, we think, must be applied to adapting any problem of 'classic' Computational Geometry to surfaces.

In the case of the width the main difficulty is that of finding an adequate definition. This happens because there are no parallel lines on the sphere, so there exist at least two possible translations of the phrase 'pair of parallel lines' to the sphere, one dropping parallel and the other omitting lines (of course, in both cases an additional condition must be considered) and the equivalence between both definitions must be proved. Once we have solved the problem of the definition, computing the width of a set will be a straightforward adaptation of one of the planar algorithms.

On the other hand, in the case of the diameter we will see that the convex hull is not a useful preprocessing, thus we will have to use another preprocessing, in this case the furthest point Voronoi diagram.

Finally, it must be pointed out that the circumradius (and some other similar problems) on the sphere has been previously treated and solved in optimal time by other authors [Hurtado et al., 2000], so in this chapter we will only summarize their results.

In this chapter we consider the width of a convex set on the sphere (the only interesting case, as we will see later) and the diameter of a set of points. Regarding the width, the relationship between some alternative definitions of the concept of the width of a convex set on the sphere is studied. Those relations allow us to characterize whether a convex set on the sphere can pass through a spherical interval by rigid motions, and lead to an optimal algorithm to compute the width on the sphere. In the case of the diameter we present an algorithm that computes the diameter of a set of n points in any surface using the furthest point Voronoi diagram as a fundamental tool. In the case of the cylinder, the sphere, or the torus, our algorithm is optimal. Finally, a combinatorial question related to the other problems of this chapter is considered. What is the maximum number of times that the largest (smallest) distance occurs amongst N points in a surface?

2. THE WIDTH OF A CONVEX SET ON THE SPHERE

The sofa problem is very important from a practical point of view, it is a major question in motion planning, and that justifies the broad literature on that subject (see, for instance, [Goldberg, 1969, Howden, 1968, Maruyama, 1973, Moser, 1966, Sebastian, 1970, Toussaint, 1985]). Basically the question is that of when is it possible to move a sofa out of an apartment, and how (see Figure 5.1).

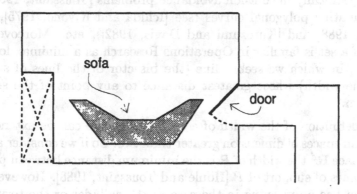

Figure 5.1. The sofa problem consists in moving a sofa out of a room and through a doorway.

The first answer to this problem appeared in [Strang, 1982] where it was proved that the width of a convex set in the plane is equivalent to the concept of the *door* of the set. The door of a set is the minimum closed interval such that the set can pass through it by a continuous family of rigid motions (translations combined with rotations), and the *width* of a set is the minimum distance between parallel lines of support of the set [Houle and Toussaint, 1988] (a line l is a line of support of a set S if all the set lies in one of the two halfplanes defined by l); i.e., a convex set can pass through a door if and only if its width is smaller than the door. Nevertheless, in dimension three this is not true and H. Stark has constructed convex sets which can pass through a door, either square or circular, although no projection of the set will fit in the doorway (see [Strang, 1982]). Although, in principle, it is not clear if the sphere shares with that of the plane the property that the width agrees with the door, we will see that, with regard to this problem, the behavior of the sphere is exactly the same as in the plane. On this point we must remark that, given the metrics considered for the cylinder and the torus, the equivalent problems on both surfaces (moving a convex body on a cylinder or a torus) are mere reformulations of the planar case. Within the sphere we can define the *circumradius* of a point set as the radius of its minimum enclosing cap, i.e., the minimum cap in the sphere containing the set.

In a similar way it is possible to define the *width of a finite set of points* in the plane as the minimum distance between parallel lines of support of the set. In addition, the concept and the computation of the width of a finite set of points have applications in several fields, such as robotics (more specifically in collision avoidance problems [Toussaint, 1985]), in approximating polygonal curves (see [Ichida and Kiyono, 1975], [Imai and Iri, 1988] and [Kurozumi and Davis, 1982]), etc. Moreover, the width of a set is familiar in Operations Research as a minimax location problem, in which we seek a line (the bisector of the lines of support given the width) whose greatest distance to any point of the set is a minimum.

The definition of the width of a set in the plane can be extended to Euclidean spaces of dimension greater than two. So if we consider a set P in the space $\mathbf{R}^d$ the width of P is the minimum distance between parallel hyperplanes of support of P [Houle and Toussaint, 1988]. However, this extension has no meaning in the case of the cylinder or the torus, but the reasons are different if we consider finite sets of points or regions delimited by a polygon. In the first case, for most point sets it is possible to find a pair of parallel helices as close as we want enclosing the set (see Figure 5.2). And in the second case the solution is exactly the same as in the plane.

In fact, at a first glance the definition is even more complicated than might be thought, since, as has been pointed out before, a helix does not divide the cylinder into two different components.

As the width of a set in the plane is the width of its convex hull [Houle and Toussaint, 1988], many authors have studied the width of convex polygons. By using the rotating caliper technique [Preparata and Shamos, 1985, Shamos, 1978] or geometric transforms [Brown, 1980] (in fact, both methods are adaptations of algorithms used to compute the diameter) it is possible to find the width of a convex polygon in linear time and space (See [Houle and Toussaint, 1988]).

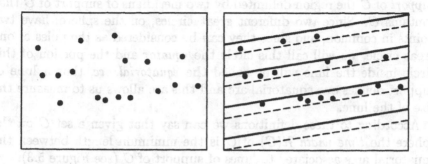

Figure 5.2. All the points of the set are between the two helices (one of the helices is indicated by solid lines and the other by dashed lines).

The structure of this section is as follows. Firstly, we will give two alternative definitions of the concept of the width of a set on the sphere and the relationship between both definitions and the width of set of points and the width of its convex hull. Then, we will seek necessary and sufficient conditions to a convex set to be able to pass through a spherical interval by rigid motions on this surface, and the relationships between the concepts described above. These conditions and relationships will allow us to adapt the rotating caliper technique to design an optimal algorithm for computing the width of a convex set in the sphere.

2.1   ALTERNATIVE DEFINITIONS OF WIDTH ON THE SPHERE

Recall that the width of a set of points in the plane is the minimum distance between parallel lines of support of the set. Trying to carry over this concept to the sphere, the main difficulty is in translating the expression 'parallel lines of support', so if we try to preserve the concept of lines (of support) we have to translate 'lines' by geodesics or great circles in the case of the sphere, but great circles in the sphere are no longer parallel (they intersect in two points). The other possible alternative is to keep the idea of parallel lines, and so we have to consider two parallels in the sphere.

So, firstly, we would replace the idea of lines of support of a set by geodesics of support of a set. In this way, given a set C on the sphere we will call *meridians of support* of C the great circles which intersect C and leave the set on one hemisphere. And now, we will call *lune of support* of C the region delimited by two meridians of support of C that contains C. Since two different great circles, on the sphere, have two points in common, and since they can be considered as the poles of one great circle, we will call this circle the *equator* and the portion of this circle inside the lune will be called the *equatorial arc*, thus a lune of support defines one equatorial arc and this arc allows us to measure the size of the lune.

According to these definitions we can say that given a set C on the sphere the *time width* $\mathcal{H}(C)$ of C is the minimum length between the equatorial arcs associated to lunes of support of C (see Figure 5.3).

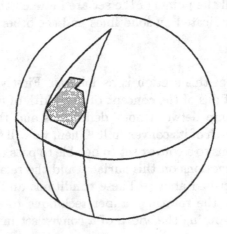

Figure 5.3.   A lune of support of a set and its equatorial arc in bold (this lune does not give the time width of the set).

As has been pointed out before, the main difference between this definition and the definition of width in the plane is that, in the plane, two parallel lines of support have empty intersection, whereas on the sphere two meridians of support have two points in common.

On the other hand, this definition has good properties for sets in Euclidean position.

LEMMA 5.1   *Given a set S in one hemisphere then $\mathcal{H}(S) = \mathcal{H}(C_m(S))$, where $C_m(S)$ denotes its m-convex hull.*

Proof: The proof of this lemma runs parallel to the proof of the same result in the plane, taking into account that a lune is always a convex set.   □

On the other hand, if we want to preserve the property that the arcs of support of a convex set have empty intersection, as in the plane, we could give another possible definition. Given a set C on the sphere, we will call a *parallel of support* of C a parallel which intersects C and leaves the set on one cap, where a *cap* is a part of the sphere divided by this parallel. If we use the idea of a pair of parallels we will conserve the concept of parallelism that we had in the plane (in the sense that they have empty intersection), but note that parallels in the sphere are not geodesics. In any case, it seems natural to ensure that some geodesics must be related to a pair of parallels. Thus we say that two parallels of support of C are a *pair of parallels of support of C* if the bisector plane of those two parallels cuts the sphere in a great circle (alternatively, they are at the same distance from their only disjoint great circle— Figure 5.4—. That circle will be called the *equator* associated with the parallels).

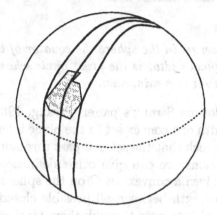

Figure 5.4.   A pair of parallels of support and its equator.

According to the definition above we can say that given a set C on the sphere the *tropical width* $\mathcal{T}(C)$ of C is the minimum distance between all possible pairs of of C.

Observe that with this definition the tropical width of a set in the sphere is not, in general, the tropical width of its convex hull (to see this point, it is enough to consider four points at the same distance — latitude— from the equator, two in the northern hemisphere and two in the southern hemisphere —see Figure 5.5).

But on the other hand it is very easy to see that the tropical width can be used, as in the plane, in Operations Research as a minimax location problem.

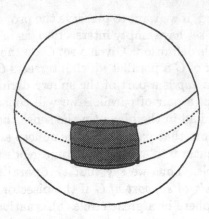

Figure 5.5.   The tropical width of a set on the sphere is not, in general, the tropical width of its convex hull.

LEMMA 5.2 *For any set in the sphere the equator of the parallels of support, given the tropical width, is the great circle whose greatest distance to any point of the set is a minimum.*

Finally, and following Strang's paper [Strang, 1982] in which it was proved that the width of a convex set in the plane is the minimum length of a closed interval such that the set can pass through it by a continuous family of rigid motions, we can give other definitions of width on the sphere as follows: given a convex set C on the sphere, the *door* $\mathcal{P}(C)$ of C is the minimum length between all possible closed arcs of meridians for the set C to be able to pass through them by a continuous family of rigid motions (it is well known that all rigid motions in the sphere are rotations such that their axis pass through the poles of a great circle) on the sphere.

G. Strang proved that the width of a convex set coincides with its door in the plane. But, as was pointed out in the introduction of this section, in dimension three this is not true, and H. Stark has constructed convex sets which can pass through a door, either square or circular, although no projection of the set will fit in the doorway (see [Strang, 1982]). Thus, it is interesting to ask if the behavior of the sphere is, in this point, similar to the plane or to the three-dimensional space.

Summarizing, given a set S on the sphere we have three invariants associated with it, its time width $\mathcal{H}(S)$, its tropical width $\mathcal{T}(S)$, and its door $\mathcal{P}(S)$. We want to relate these three invariants, and the first easy answer is given by the following two lemmas.

LEMMA 5.3 *Let C be a convex set on the sphere. Then* $\mathcal{P}(C) \leq \mathcal{T}(C)$.

Proof: It suffices to consider the arc of meridians orthogonal and contained between the parallels which define $\mathcal{T}(C)$. The length of this arc is greater than or equal to $\mathcal{P}(C)$, and, obviously, less than or equal to $\mathcal{T}(C)$. □

LEMMA 5.4 *Let C be a convex set on the sphere. Then* $\mathcal{T}(C) \leq \mathcal{H}(C)$.

Proof: Let $\mathcal{H}$ be the lune that defines $\mathcal{H}(C)$. This lune is defined by meridian arcs which intersect C in two points P and Q. We consider the parallels tangent to C at the points P and Q. The distance $\mathcal{T}^*$ between these parallels is equal to $\mathcal{H}(C)$, so $\mathcal{T}(C) \leq \mathcal{T}^* = \mathcal{H}(C)$. □

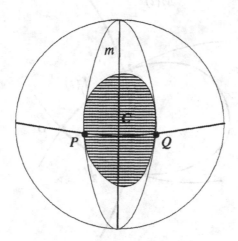

Figure 5.6.   Illustration of the proof of Lemma 5.4.

Therefore we already have establish that $\mathcal{P}(C) \leq \mathcal{T}(C) \leq \mathcal{H}(C)$. The next theorem says that, as happens in the plane, these three numbers agree on the sphere.

THEOREM 5.1 *A convex set C on the sphere can pass through a meridian arc of length* $\mathcal{L}$ *if and only if* $\mathcal{H}(C) \leq \mathcal{L}$. *As a consequence* $\mathcal{P}(C) = \mathcal{T}(C) = \mathcal{H}(C)$.

Proof: If $\mathcal{H}(C) \leq \mathcal{L}$ it is clear that the lune of length $\mathcal{H}(C)$ can pass through a door of length $\mathcal{L}$ and the set C is contained in that lune.

To prove the converse, assume first that the boundary ∂C of C is smooth, this is to say, through every boundary point there is a unique tangent line on the sphere, and it varies continuously along ∂C. Let I be an arc of meridian of length $\mathcal{P}(C)$ and denote the sphere by S. As C can pass through I it is possible to define a continuous composition of motions $M : [0,1] \to S$, where $M(0)$ is the location of C at the starting point before crossing I and $M(1)$ is the situation after passing through I. For all $t \in [0,1]$ we can define two applications $f_1 : [0,1] \to [0,\pi]$ and $f_2 : [0,1] \to [0,\pi]$ as follows: $f_1(t)$ and $f_2(t)$ are the angular lengths between the points P_1 and P and the points P_2 and P, respectively, where P_1 and P_2 are the intersection of ∂C with the arc I and P is the intersection between the tangents to C at the points P_1 and P_2 (see Figure 5.7).

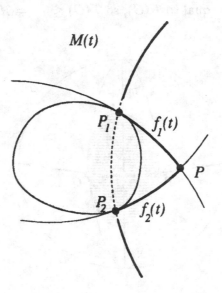

Figure 5.7. Illustration of the applications defined in the proof of Theorem 5.1.

The map $f_1 + f_2 : [0,1] \to [0,2\pi]$ is continuous and at the extremities of the interval $[0,1]$ we know that $f_1(0) + f_2(0) = 0$ and $f_1(1) + f_2(1) = 2\pi$, so there exists $t^* \in [0,1]$ such that $f_1(t^*) + f_2(t^*) = \pi$.

If $f_1(t^*) = f_2(t^*) = \frac{\pi}{2}$, then the meridian arc which defines $\mathcal{H}(C)$ is contained in I, so $\mathcal{H}(C) \leq \mathcal{L}$. Otherwise, $f_1(t^*) - \frac{\pi}{2} = \frac{\pi}{2} - f_2(t^*)$. Suppose that $f_1(t^*) > \frac{\pi}{2}$ and so $f_2(t^*) < \frac{\pi}{2}$. The situation is then as in Figure 5.8.

As the angles in the equatorial arc of the lune considered are of ninety degrees, the length of the meridian arc joining P_1 and P_2 is greater than or equal to the length of that meridian arc. So $\mathcal{H}(C) \leq \mathcal{L}$.

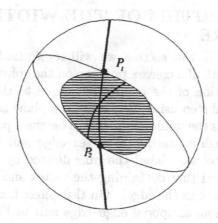

Figure 5.8. The situation when $f_1(t^*) > \frac{\pi}{2}$.

The conclusion remains true for a convex set C even if ∂C is not smooth. We will proceed by introducing a sequence of smooth convex subsets C_n converging to C. As C passes through I so do the C_n, and their time widths must satisfy $\mathcal{H}(C_n) < \mathcal{L}$. Therefore $\mathcal{H}(C) \leq \mathcal{L}$ and the theorem is proved. $\qquad\square$

The three definitions of width we have considered previously then agree in the case of convex sets, and we can talk about the width of a convex set.

As a consequence of Theorem 5.1 we obtain:

COROLLARY 5.1 *The equatorial arc of the lune giving the width of a convex set C is included in C.*

Proof: Consider an interval I of length $\mathcal{L} = \mathcal{H}(C)$. By Theorem 5.1 we know that it is possible to move C through that interval, but if the width of C agrees with the length of the interval, at some instant, two points of C, say P and Q, must simultaneously touch the extremes of I (otherwise it is possible to find a shorter interval such that C can cross it). By definition P and Q are the extremes of the equatorial arc defining the time width of C, and by convexity that equatorial arc must be contained in C. $\qquad\square$

2.2   ALGORITHM OF THE WIDTH ON THE SPHERE

In what follows in this section we will try to find an algorithm for computing the width of a convex polygon on the sphere. This algorithm will be an adaptation of the caliper technique to this surface. Recall that a fundamental step using calipers in the plane is that the width of a convex polygon is the minimum distance between parallel lines of support passing through an antipodal vertex–edge pair (to each antipodal vertex–edge pair, we associated the lune defined by the meridian containing the edge and that containing the vertex such that the equator arc joins the vertex with the edge). On the sphere this is not true and it can be achieved in an antipodal edge–edge pair as Figure 3 shows, but we have:

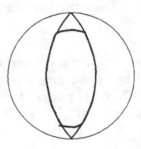

Figure 5.9.   The time width of a set can be achieved in an antipodal edge–edge pair.

LEMMA 5.5 *The width of a convex polygon is the minimum distance between meridians of support passing through either an antipodal vertex–edge pair or an edge–edge pair.*

Proof: Let H be the minimum lune containing the convex polygon C. By Corollary 5.1 at least two points of the equatorial arc of H must be in C. But at least one of the meridians defining the lune must contain an edge of C, otherwise it would be possible to find a smaller lune.   □

In any case, Lemma 5.5 tells us that it is possible to adapt the rotating caliper algorithm for finding the width of a convex polygon C as follows:

- Find an antipodal pair in the following way. Consider an edge a of C and a vertex v other than the extremes of a. The minimum arc joining v with the great circle which contains a defines a lune (that lune must contain a). If that lune contains C, a and v are antipodal, otherwise if a_1 and a_2 are the edges containing v, by the convexity of C either $a - a_1$ or $a - a_2$ is an antipodal pair. Now we compute the equatorial arc associated with the antipodal pair so obtained. This is to say, each antipodal pair defines a lune and we compute the equatorial arc of that lune.

- Rotate the meridians in the same way as the calipers in [Houle and Toussaint, 1988], until finding the next antipodal pair, and so on.

- Give the minimum arc of the lunes obtained in previous steps and containing the set as the width of C.

Thus we can give the algorithm described in Table 5.1.

ALGORITHM WIDTH(C)

Input: *Convex polygon C on the sphere.*

Output: *Width of C.*

Step 1 *Find an initial vertex-edge pair.*

Step 2 *If the lune associated with the vertex–edge pair contains C, compute its equator arc, otherwise compute the equator arc of the pair edge–edge associated with the original vertex–edge pair (this edge–edge pair is defined from the vertex–edge pair by considering the edge incident with the vertex that is not contained in the lune).*

Step 3 *Use the rotating caliper technique to generate all pairs as in (1)–(2).*

Step 4 *Compute the minimum obtained in previous steps.*

Table 5.1. Algorithm for computing the width of a convex set on the sphere.

Therefore:

THEOREM 5.2 *Algorithm* WIDTH(C) *computes the width of a convex polygon C in optimal linear time.*

Proof: The validity of the algorithm WIDTH(C) follows from the same arguments used for proving the validity of the rotating calipers technique used in the plane for computing the width of convex polygons. Finally, the algorithm is linear because the number of events considered is the number of rotations of meridians that agrees with the number of antipodal pairs that are lineal.   □

3.   CIRCUMRADIUS

In some sense the circumradius is closely related to one of the key concepts in this book, that of Euclidean position, but it has been considered generally as a minimax location problem. As we have said in the Introduction of this chapter, the *circumradius* is the radius of the minimum cap enclosing a set of points on the sphere. Obviously if this functional is smaller than $\pi/2$ then the set is in Euclidean position. But, on the other hand, the center of the minimum enclosing cap is a point that minimizes the maximum distance to the set of points.

This problem has been studied in [Hurtado et al., 2000] and in [Sacristán, 1997], where they obtain that solving this problem requires, in general, $\Omega(n \log n)$ operations, but that if the sites are on an open hemisphere (in other words, if the set is in Euclidean position), it is possible to find the circumradius in linear time.

4.   DIAMETER

A well known measure of the spread of a set is its *diameter* (i.e., the maximum distance between two points of the set). Intuitively speaking, a cluster with small diameter has elements that are closely related, while the opposite is true when the diameter is large. So the diameter of a point set is an important parameter, used, for instance, in size normalization. This concept has led to several related problems producing a remarkable amount of literature (see, for instance, [Akl, 1979, Freeeman, 1974, Freeman and Shapira, 1975, Rosenfeld, 1969, Sklansky, 1972]). But most of the efforts have been concentrated in the plane or Euclidean spaces, and, in many cases the set of points in which we are interested are not in an Euclidean space but confined to some surface (or a more general space) and the usual techniques are no longer valid.

It is known that the computation of the diameter of a set of N points in every Euclidean space requires $\Omega(N \log N)$ operations (see, for example, [Preparata and Shamos, 1985]). The usual procedure for computing in optimal time the diameter in the plane uses the property that the diameter of a set of points is equal to the diameter of its convex hull ([Hocking and Young, 1961]), then it is enough to compute all antipodal pairs, and, in a convex polygon, this task can be completed in linear time (using rotating calipers, for instance), thus the total running time of the algorithm is $O(N \log N)$. Unfortunately, this method cannot be used in the space since the number of antipodal pairs in the space is $O(N^2)$. And, in fact, an $O(N \log N)$ algorithm in dimension 3 is not known (as far as we know, the best result for the running time of a deterministic algorithm for the three-dimensional diameter problem is an $O(N \log^3 N)$ algorithm due to Amato, Goodrich and Ramos [Amato et al., 1994]).

In this section we will show that although the procedure followed in the plane for computing the diameter cannot be applied to many surfaces (the cylinder is one of those surface), it is possible to obtain an $O(N \log N)$ time algorithm on those surfaces by using the furthest point Voronoi diagram. Finally, if we can compute that structure in $O(N \log N)$ time (and we can as we have seen in Chapter 4), we will have an optimal algorithm for computing the diameter of a point set.

Obviously the main obstacle on the cylinder, as on the other surfaces, is that the convex hull of a set of points is, in general, too big (see Chapter 3), and therefore, it is not a useful tool for most problems. In fact, it is not difficult to find examples of sets of points on the cylinder such that their diameters are not equal to the diameters of their convex hulls (see Figure 5.10. Therefore, another technique is needed in order to obtain an optimal algorithm.

The same situation happens on the torus and on the sphere. Of course, if the points are in Euclidean position the behavior of that point set is exactly as in the plane, therefore the convex hull does determine the diameter and so we will consider only point sets that are not in Euclidean position.

In fact, the key of our algorithm is a very easy observation:

LEMMA 5.6 *Let S be a point set. If the diameter of S is obtained for $u, v \in S$ then $u \in Vor^F(v)$.*

Hence an algorithm for computing the diameter will be that presented in Table 5.2.

It is straightforward to check the validity of the algorithm DIAMETER(S) and then we have

THEOREM 5.3 *It is possible to compute the diameter of a set of N points on the cylinder (or the torus, or the sphere) in optimal $O(N \log N)$ time.*

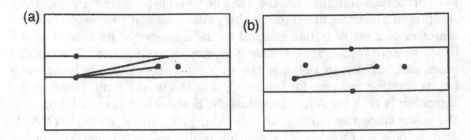

Figure 5.10. The diameter of a point set in the cylinder is not equal to the diameter of its convex hull. In (a) we can see that the diameter of a point set can be smaller than the diameter of its convex hull and in (b) we can see that the diameter of the point set can be achieved by points that are not extreme points.

ALGORITHM DIAMETER(S)

    Input: *A set of points S.*

    Output: *Diameter of S.*

    Step 1 *Construct the furthest point Voronoi diagram of S.*

    Step 2 *Localize in which region of the diagram is each point of S.*

    Step 3 *Compute the distance between each point of S and the point defining the region obtained in Step 2.*

    Step 4 *Report the maximum obtained in Step 3 as the diameter.*

Table 5.2. Algorithm for computing the diameter of a set of points

Proof: As we have said before, the validity of the theorem is an easy consequence of Lemma 5.6. Regarding the complexity of DIAMETER(S), we know by Theorem 4.2 that Step (2) can be performed in $O(N \log N)$ time. So the knowledge that Step (3) (to localize a point in a given subdivision of our surfaces) can be performed in $O(\log N)$ for each point (for a total of N points) completes the proof. $\qquad\square$

5.   MAXIMUM AND MINIMUM DISTANCES

An interesting question which has been studied extensively in Euclidean spaces is that of, how many times can the maximum distance between N points occur?

It is known that in the plane it can occur at most N times [Erdos, 1946], and in the space $2N - 2$ times [B. Grünbaum, 1956]. The case of our surfaces is not quite the same as in the plane, since it is possible to give an N-point set on the cylinder, or on the torus on each of which the maximum occurs $4/3N$ times. In fact, the case of the sphere agrees with the 3-dimensional space rather than with the plane since it is possible to reproduce the critical configuration on the sphere, this can be done by placing $N - 1$ points forming a regular polygon on an adequate parallel and the other point at the north pole as Figure 5.11 shows.

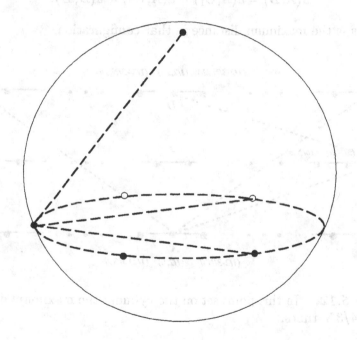

Figure 5.11.   In this point set on the sphere the maximum distance occurs $2N - 2$ times.

Regarding the cylinder and the torus, as we have commented above, it is possible to find a structure with N points such that the maximum is reached $4/3N$ times. This structure split the N-point set into three subsets of $N/3$ points each and each of those subsets forms a regular polygon on a parallel in such a way that the polygons in the top and in the bottom parallels have their vertices on the same meridians and the other polygon has its vertices on the meridians equidistant to those considered before, see Figure 5.12. More concretely, consider $b > 0$ and $k = N/3$, the set S is $S = \{(\frac{t}{k}, 0) : t = 0, \ldots, k-1\} \cup \{(\frac{t}{k} + \frac{1}{2k}, b) : t = 0, \ldots, k-1\} \cup \{(\frac{t}{k}, 2b) : t = 0, \ldots, k-1\}$. Now, if we fix

$$b = \sqrt{\frac{1}{3}(\frac{1}{4} - \frac{(N-3)^2}{4N^2})},$$

it is easy to see that for the points labeled as in Figure 5.12,

$$d(A, B) = d(A, C_1) = d(A, C_2) = d(B, D),$$

and this is the maximum distance in that configuration.

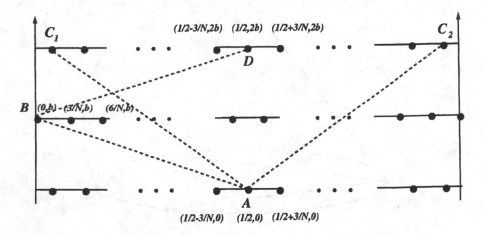

Figure 5.12.   In this point set on the cylinder the maximum distance occurs $4/3N$ times.

It is an open problem whether this example is optimal or whether it is possible to find better bounds.

On the other hand, it is possible to answer completely a related question, how many times can the minimum distance between N points occur? In the plane this problem is more difficult that the equivalent problem of the maximum distance, and it is known that this value is

$\lfloor 3N - \sqrt{12N - 3} \rfloor$ ([Harborth, 1974, Moser and Pach, 1993]). However, on the cylinder it is an easier problem.

THEOREM 5.4 *The minimum distance between N points on the cylinder can occur at most $3N - 6$ times, and this bound is tight (see Figure 5.13).*

Proof: Given a set of N points on the cylinder, if we join two points by a segment if that couple gives the minimum distance, then there are no crossings between those segments (otherwise, if two segments cross each other, the extremes define a quadrilateral, and at least one of the sides of the quadrilateral is shorter than the diagonals which are the original segments given the minimum distance). Hence the graph so constructed can be embedded in the cylinder and therefore it is planar. Thus by Euler's formula $3N - 6$ is an upperbound of the number of edges of that graph, and this gives the upperbound. Finally, Figure 5.13 shows that this upperbound can be achieved.   □

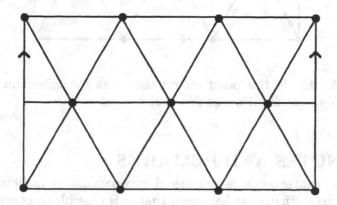

Figure 5.13. In this point set on the cylinder the minimum distance occurs $3N - 6$ times (for this case $N = 9$).

Finally, observe that this kind of construction can be translated to the torus obtaining, in this way, tight bounds (note that we have to use Euler's formula for the case of the torus).

THEOREM 5.5 *The minimum distance between N points on the torus can occur at most $3N$ times, and this bound is tight (see Figure 5.14).*

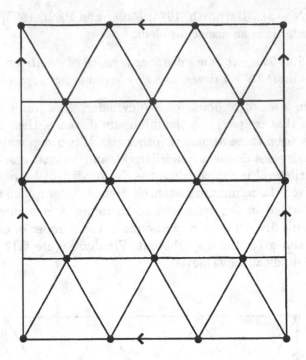

Figure 5.14.   In this point set on the torus the minimum distance occurs $3N$ times (for this case $N = 12$).

6.   NOTES AND REMARKS

In this chapter we have considered two parameters associated with sets of points. Firstly, we have seen that it is possible to generalize the concept of the width of a set on the plane to the sphere. In fact, several definitions of width on the sphere arise, depending whether we want to preserve the condition of geodesics in the plane (lines) for geodesic on the sphere, or whether we want to preserve the parallelism property, or whether we want generalize the property of passing through a slot in the plane to the property of passing through a meridian arc on the sphere. In the case of convex sets all those definitions agree and this property allows us to adapt in a straightforward way the corresponding algorithm for computing the width of a set to the case of the sphere.

Regarding the other parameter, the diameter of a point set, although an optimal algorithm for computing this invariant is presented (based on the furthest point Voronoi diagram), some open questions arise related to this problem.

First of all, it would be interesting to find a structure that, as does the convex hull in the plane, allows us to find from it the diameter on the cylinder or some other surface in linear time (observe that from the furthest point Voronoi diagram we find the diameter in $O(N \log N)$ time). It is not clear whether this structure could be the minimum enclosing polygon considered in Chapter 2.

In addition we propose here what we think is an interesting combinatorial question, namely, how many times can the maximum distance occur? we do not know the exact answer to this question, but it seems to be something in between the plane and the space. However, we know that the minimum distance can occur on the cylinder at most $3n - 6$ times.

Most of the results of this chapter (although not all the details) have appeared previously in [Dana et al., 1998, Cobos et al., 1997c, Cobos et al., 1997a, Cobos et al., 1997b, Cobos et al., 1997d].

Chapter 6

VISIBILITY

Visibility questions constitute a prolific subfield in Computational Geometry. In a geometric context two object are visible to each other if there is a line segment connecting them which does not cross any obstacles. Therefore as visibility is associated with geodesics, when considering these problems on surfaces important differences arise with regard to the plane or space (more than one geodesic can join two points). To show these differences and how to approach them is the main objective of this chapter.

We will not treat here classical Art Gallery Theorems (see [Goodman and O'Rourke, 1997, O'Rourke., 1987]) but some questions related to visibility amongst obstacles, the visibility graph, and, mainly, a closely related problem such as the stabbing line problem.

1. INTRODUCTION

Probably more than any other problem in Computational Geometry, visibility problems have been studied only in an Euclidean context. The main reason for this fact is that visibility, in general, is associated with some applications to Computer Vision, Computer graphics, etc., which usually are constrained to the plane or the 3-dimensional space. Nevertheless, some authors have considered non-Euclidean visibilities, such as the circular visibility treated in [Agarwal and Sharir, 1993].

The stabbing line problem [Edelsbrunner, 1987, Edelsbrunner et al., 1982] is considered here from two points of view. For a given segment set we stab them either with geodesics or with bisectors, obtaining two completely different problems (in the case of the cylinder or the torus).

107

Complexities in both cases are very interesting, because with geodesics we obtain a complexity which depends on the complexity of the input data, and with bisectors the complexity is higher.

On the other hand, if we consider, as we said above, that two points in the presence of obstacles can see each other if there exists a geodesic joining them which does not intersect any obstacle. We can see that because infinitely many geodesics join two points there is an important difference with respect to the same question in the plane. But the techniques used for solving the problem of whether a set of segments admits a transversal helix will allow us to solve this visibility problem. In fact, this is a particular case of the construction of the visibility graph. It must be remarked that this structure is, in general, infinite, so we will discuss a method of recognizing the finiteness of the visibility graph.

The structure of this chapter is as follows. In Section 2 we will study the two variants mentioned above, so firstly we will try to find a transversal helix. And, secondly, we will stab segments with geodesics. Finally, we will use the results obtained in Section 2, specially those considered when we try to find a transversal helix, for studying when two points can see each other in the cylinder in the presence of some obstacles.

2.   STABBING LINE SEGMENTS

In this section the classical problem in Computational Geometry of stabbing line segments [Edelsbrunner, 1987, Goodman and O'Rourke, 1997] is revisited, considering segments on the cylinder instead of line segments in the plane. This problem is interesting both from a combinatorial and from a computational point of view. And translating it to the cylinder we obtain important variations with respect to the results obtained in the plane or in the space.

We recall that given a set S of N not necessarily disjoint line segments in the plane, a straight line t is called a *transversal* of S if it intersects each segment in S. We also say that t stabs S.

Obviously the generic problem of stabbing line segments is to find transversals for given sets. This problem, and many variations, has been studied in the plane for sets of segments ([Edelsbrunner et al., 1982, O'Rourke, 1981, Katchalski et al., 1985]); in addition, it is possible to find algorithms computing transversals for rectangles, circles, or polygons ([Atallah and Bajaj, 1987, Avis and Wenger, 1988, Edelsbrunner, 1985, Jaromczyk and Kowaluk, 1988]). Finally, we want to remark that the importance of this problem can be illustrated if we think that in the book [Edelsbrunner, 1987] in order to obtain concrete examples of

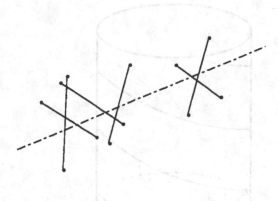

Figure 6.1. The dashed line is a transversal for a set of segments in the plane.

design paradigms the author applies each one to a variant of a transversal problem defined for line segments in the plane.

As in the cases of the diameter or the width of point sets, the first and crucial task will be to obtain right definitions in the cylinder. Thus again, as in the case of the width, two possibilities arise. Firstly, as the lines are the geodesics of the plane, we can try to find a transversal amongst the helices of the cylinder. But to generalize straight lines as geodesics presents the inconvenience that those curves do not split the cylinder into two parts (see Figure 6.2).

Thus if we seek a line splitting the cylinder into two parts we find that straight lines are the bisectors of the plane. Bisectors in the cylinder were considered in Chapter 3, where we showed some of their properties. Of course, the first property which we will use in this chapter is that they split the cylinder into two parts. A second property is that, besides them being defined as the locus of points on the cylinder to the same distance from two points, they share another property with planar straight lines: they are defined by two of their points, that is, there exist two points in each bisector such that the whole bisector can be constructed from them simply considering the union of the two geodesics with least length joining those points (Lemma 4.2).

In this way two possible characterizations of straight lines are considered here. On the one hand, if we think of straight lines as the geodesics in the plane, we will think of the helices of the cylinder (its geodesics). On the other hand, we can refer to the straight lines as the bisectors between two points of the plane, and in this case helices are not appro-

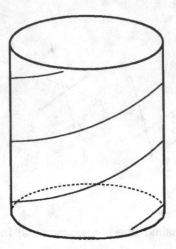

Figure 6.2.   A helix does not split the cylinder into two parts (its complement has only one connected component.

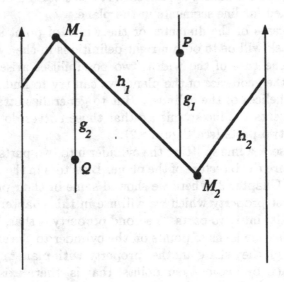

Figure 6.3.   The bisector of the points P and Q is the union of the two helices with least length joining M_1 and M_2 (see text for an explanation on how to obtain M_1 and M_2).

priate generalizations, but the bisectors we have just described above. Therefore the two possible generalizations lead to two different definitions in the cylinder. Thus we say that if S is a set of n segments on the cylinder a helix is a *transversal* of S if it intersects each segment in S. And we will say that a bisector b *stabs* S if each segment in S has an extreme in each part of the cylinder defined by b (note that a bisector might not stab all the segments that it intersects —see Figure 6.4).

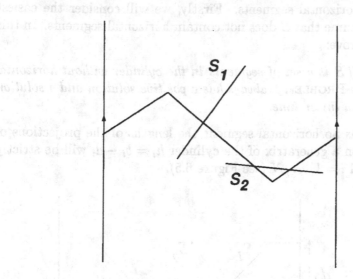

Figure 6.4. The bisector of this figure stabs the segment s_1 but it only intersects the segment s_2.

So the main purpose of this section is to give conditions on a set of segments on the cylinder to ensure that it admits a transversal or a stabbing bisector.

2.1 TRANSVERSAL HELICES

In the first case, when we use helices we face the following natural problem:

PROBLEM 1. Determine whether or not a given finite set of segments on the cylinder admits a transversal.

This problem seems to be very similar to the equivalent problem in the plane, but we will see that its answer is quite different. In order to solve PROBLEM 1 we will consider the representation of the cylinder in bands, and we will start the helices in a fixed band (see Chapter 1). Let $S = \{s_1, s_2, \ldots, s_N\}$ be a set of segments on the cylinder, and let $p_i = (a_i, b_i)$ and $q_i = (c_i, d_i)$ be the extremes of segment s_i satisfying $b_i \geq d_i$ (p_i is the highest extreme of s_i).

We will see that the most important property in this problem is the existence of horizontal segments. Firstly, we will consider the easiest case, so we assume that S does not contain horizontal segments. In this case we can prove:

LEMMA 6.1 *If S is a set of segments in the cylinder without horizontal segments, then* PROBLEM 1 *always has a positive solution and a solution can be found in linear time.*

Proof: If S has no horizontal segment the lengths of the projections of all segments on a generatrix of the cylinder $h_i = b_i - d_i$ will be strictly positive for all $i = 1, \ldots, N$ (see Figure 6.5).

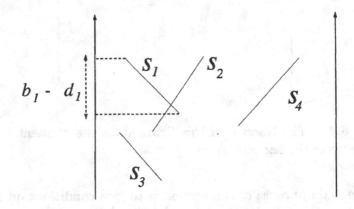

Figure 6.5.   The projection of a segment on a generatrix of the cylinder.

Let $h = \min\limits_{i=1,\ldots,N} h_i$; if we consider a helix H with pitch $h/2$ that helix is a transversal for S. To show this last claim we can divide the developed cylinder into horizontal bands or rectangles of height $h/2$, the diagonal of each one of those rectangles will be the helix H, as is shown in Figure 6.6

Now let s_k be a segment in S. If we consider the rectangle that contains q_k in its bottom side, and p_k in its top side as Figure 6.7 shows, since $h_k \geq h$ that rectangle must contain completely whithin its interior one of the rectangles of height $h/2$ defined above, and therefore the diagonal of this last rectangle (that it is a portion of H) intersects s_k.

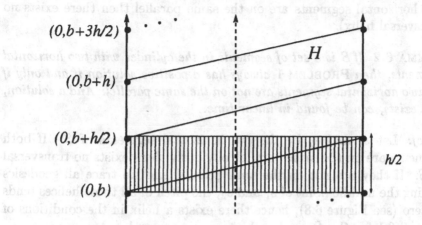

Figure 6.6. The helix H is the diagonal of the rectangles of height $h/2$ on the cylinder.

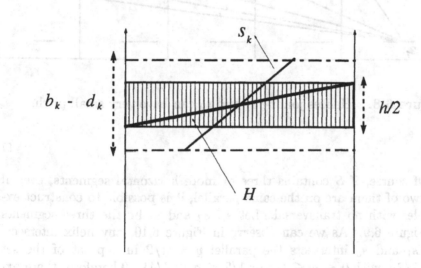

Figure 6.7. The helix H intersects the segment s_k.

□

Of course, if S contains only one horizontal segment the result of Lemma 6.1 is still valid. Even if S contains two horizontal segments that are not on the same parallel we can obtain the same result (if the two horizontal segments are on the same parallel then there exists no transversal helix).

LEMMA 6.2 *If S is a set of segments in the cylinder with two horizontal segments, then* PROBLEM 1 *always has a positive solution if and only if the two horizontal segments are not on the same parallel. And a solution, if it exists, can be found in linear time.*

Proof: Let s_1 and s_2 be the two horizontal segments of S. If both segments are on the same parallel, obviously there exists no transversal of S. If they are not on the same parallel and we trace all geodesics joining the middle point of s_1 and s_2, the pitch of all those helices tends to zero (see Figure 6.8), hence there exists a helix in the conditions of Lemma 6.1 for $S - \{s_1, s_2\}$ and intersecting s_1 and s_2.

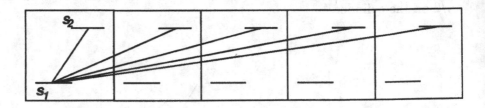

Figure 6.8. Helices joining s_1 and s_2 can have very small pitch.

□

Of course, if S contains three or more horizontal segments, even if no two of them are on the same parallel, it is possible to construct examples with no transversal. Let s_1, s_2 and s_3 be the three segments of Figure 6.9. As we can observe in Figure 6.10, any helix intersecting s_1 and s_3 intersects the parallel $y = 1/2$ in a point of the set $\{(a, 1/2);$ with $0 \leq a \leq 1/4$ or $1/2 \leq a \leq 3/4\}$. Therefore, there exists no common transversal for those three segments.

Once we know that the answer of PROBLEM 1 can be in the negative, we will describe here an algorithm that decides whether a set of segments admits a transversal, and calculates one if that is possible.

Obviously, the first step of the algorithm will be to check how many horizontal segments the initial set S has. If the answer is 'none' or 'one' or 'two', we know that S admits a transversal and we know how to calculate it. Thus without loss of generality we can assume that all segments in $S = \{s_1, \ldots, s_N\}$ are horizontal; and let $p_i = (a_i, b_i)$ and

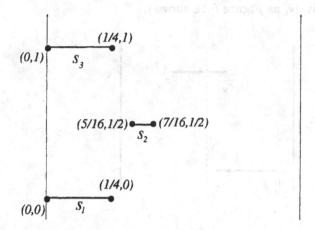

Figure 6.9. Three horizontal segments which do not admit any transversal.

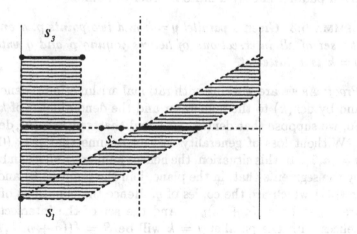

Figure 6.10. All helices joining s_1 and s_3 intersect the parallel $y = 1/2$ in the shaded region.

$q_i = (c_i, b_i)$ be the extremes of segment s_i verifying that $b_1 \leq b_2 \leq \cdots \leq b_N$. As a first step we will fix our attention on only two of them (that will be s_1 and s_N). Without loss of generality we can assume that

$a_1 \le c_1$ and that $a_N \ge c_N$ (p_1 and p_N are the left extremes of s_1 and s_N, respectively, as Figure 6.11 shows).

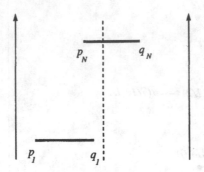

Figure 6.11.   The situation described in the text with two horizontal segments.

A key result is that all possible helices joining two points intersect a given parallel in only a finite number of points.

LEMMA 6.3 *Given a parallel $y = k$ and two points p, q on the cylinder, the set of all intersections of helices joining p and q with the parallel $y = k$ is a finite set.*

Proof: As we are working with rational arithmetic we denote by $\text{num}(k)$ and by $\text{den}(k)$ to the numerator and the denominator of k respectively. So, we suppose that $\text{den}(k) \in \mathbf{Z}^+$ and that $\gcd(\text{num}(k), \text{den}(k)) = 1$.

Without loss of generality we can assume that $p = (0,0)$ and that $q = (a, 1)$. In this situation, the helices joining p and q can be represented by the segments that, in the plane, join the point $(0,0)$ and $\{(a+m, 1) : m \in \mathbf{Z}\}$ which are the copies of q. Hence the equations of those helices are $r_m \equiv \{x = (a+m)y\}$. And the set of the intersections of those helices with the parallel $y = k$ will be $S = \{((a+m)k, k) : m \in \mathbf{Z}\}$. Of course, S is an infinite set of points on $\mathbf{R}^2$, but if we consider the quotient space in order to obtain the cylinder we obtain that, at most, $\text{den}(k)$ of them are distinct points on the cylinder.   □

The last lemma allows us to know the region in which all helices between s_1 and s_N lie. We can consider that the coordinates of the extremes of s_1 are $(0,0)$ and $(c_1, 0)$; and the coordinates of the extremes of s_N are $(a_N, 1)$ and $(c_N, 1)$

LEMMA 6.4 *Given two horizontal segments s_1, s_N on the cylinder, the set of all intersections of helices joining s_1 and s_N with the parallel $y = k$ is the union of the segments t_m for $m = 1, \ldots, den(k)$, where*

$$t_m = \{(sa_N + m)k + (1 - s)[(c_N + m - c_1)k + c_1], k) : s \in (0, 1)\}$$

Proof: This lemma is an easy consequence of Lemma 6.3. □

Observe that all segments obtained in Lemma 6.4 have the same length $(c_N - c_1 - a_N)k + c_1$. In what follows we will denote by $\mathcal{T}(s_k)$ the union of the family of segments t_m obtained in Lemma 6.4 for the parallel containing the horizontal segment s_k (see Figure 6.12).

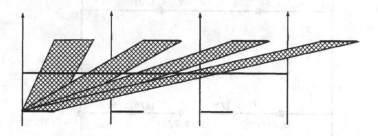

Figure 6.12.   All helices joining two segments lie in the shaded region (here we depict only those segments with positive slope).

At this point it is possible to design an algorithm for solving PROBLEM 1. Thus we can compute the intersection $s_2 \cap \mathcal{T}(s_2)$. If that intersection is empty then PROBLEM 1 has no solution for s_1, s_N and s_2 (therefore, it has no solution for the whole set S). On the other hand, if that intersection is not empty we have to compute it and to carry that information to check if that intersection is compatible with $s_3 \cap \mathcal{T}(s_3)$, and so on. We are now going to see that all of this process can be done. In order to obtain this goal we will label the segments of $\mathcal{T}(s_i)$; that labeling will tell us the ordering in which the segments of $\mathcal{T}(s_i)$ occur.

In order to obtain this goal we are going to consider all geodesics joining two points $p = (0, 0)$ and $q = (a, 1)$, where a is a rational number between 0 and 1. The equations of the helices joining p and q on the cylinder are the equations of the lines in the plane joining $p = (0, 0)$ with the copies of q ($q_m = (a + m, 1)$, $m \in \mathbf{Z}$); those equations are $h_m = \{x = (a + m)y\}$. Consider $N - 2$ parallels in the cylinder $\{y = b_2, y = b_3, \ldots, y = b_{N-1}\}$ with $b_k = \frac{n_k}{d_k}$ and $0 < b_2 < b_3 < \ldots < b_{N-1} <$

1. As we have seen above the intersection points between each parallel $y = b_k$ with all helices h_m, $m \in \mathbf{Z}$, are $\{\beta_m^k = ((a+m)b_k, b_k), m = 0, 1, 2, \ldots, d_k - 1\}$. These points divide the parallel $y = b_k$ into d_k parts with the same length. Now let $l = \mathrm{lcm}(d_2, d_3, \ldots, d_{N-1})$. We assign labels in the following way: in each parallel $y = b_k$ we call $\alpha_0^k = \beta_0^k$ and $\alpha_j^k = \beta_{n_j}^k$ with $n_j = jn_k (\mathrm{mod}.d_k)$, for $j = 1, 2, \ldots, l - 1$. In this way, in general, each point has more than one label.

In Figure 6.13 is shown this labeling process with $p = (0,0)$, $q = (1/2, 1)$ and the parallels $y_1 = 2/5$ and $y_2 = 1/2$.

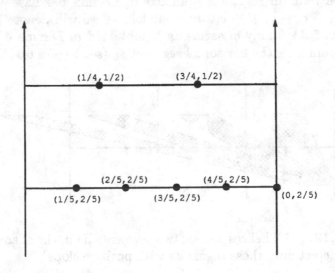

Figure 6.13.   The steps of the labeling process are the following: In a first step points are labeled as $\{\beta_0^1 = (1/5, 2/5), \beta_1^1 = (2/5, 2/5), \beta_2^1 = (3/5, 2/5), \beta_3^1 = (4/5, 2/5), \beta_4^1 = (0, 2/5)\}$; $\{\beta_0^2 = (1/4, 1/2), \beta_1^2 = (3/4, 1/2)\}$. In a second step, we get: $\{\alpha_0^1 = \alpha_5^1 = \beta_0^1 = (1/5, 2/5), \alpha_6^1 = \beta_2^1 = (3/5, 2/5), \alpha_2^1 = \alpha_7^1 = \beta_4^1 = (0, 2/5), \alpha_3^1 = \alpha_8^1 = \beta_1^1 = (2/5, 2/5), \alpha_4^1 = \alpha_9^1 = \beta_3^1 = (4/5, 2/5)\}$. And in the third step: $\{\alpha_0^2 = \alpha_2^2 = \alpha_4^2 = \alpha_6^2 = \alpha_8^2 = (1/4, 1/2), \alpha_1^2 = \alpha_3^2 = \alpha_5^2 = \alpha_7^2 = \alpha_7^2 = (3/4, 1/2)\}$.

With this labeling we know that if a helix h intersects the parallel $y = b_2$ in the point with the label α_j^2, then the other parallels $y = b_k$, $3 \le k \le N - 1$, intersect h in the points with labels α_j^k.

We will now use this process in the set of segments S. Thus in each parallel $y = b_k$ we assign to the segments of the family $\mathcal{T}(b_k)$ the labeling α_j^k applied to the intersections of the helices joining p_1 to p_N, i.e., we assign to each segment the label of its leftmost extreme. In the first step each label α_j^k represents a segment as $\{(sA_j^k + (1-s)B_j^k, b_k); s \in [0,1]\}$, that is to say that in the first step the labels of the segments will be $\alpha_j^k = [0,1]$. Once each parallel $y = b_k$, $2 \le k \le N-1$, has labeled the segments of $\mathcal{T}(b_k)$ we proceed as shown in Table 6.1.

Algorithm TRANSV-HORIZONTAL(S)

Input: $S = \{s_1, s_2, \ldots, s_N\}$ *horizontal segments in the cylinder.*

Output: *A transversal for S, if it exists.*

Step 1 *From $k = 2$ to $k = N - 1$ do*

    1.1 *From $j = 1$ to $j = l - 1$ ($l = lcm(d_2, \ldots, d_{N-1})$) do*

      a. *If $s_k \cap \alpha_j^k = \emptyset$ we eliminate the labels α_j^n $n \ge k$;*

      b. *If $s_k \cap \alpha_j^k \ne \emptyset$, then if $s_k \cap \alpha_j^k = \{(sA_j^k + (1-s)B_j^k, b_k); s \in [a_j, b_j] \subseteq [0,1]\}$, then we change the labels $\alpha_j^n = [a_j, b_j]$ for $n \ge k$.*

Step 2 *If all labels have been removed then there exists no transversal for S, END.*

Step 3 *For any label $(\alpha_j^{N-1}, [a_j, b_j])$ the straight line that joins $(0,0)$ and a point $(sA_j^{N-1} + (1-s)B_j^{N-1}, b_{N-1})$, for $s \in [a_j, b_j]$, will represent a transversal helix for S.*

Table 6.1. Algorithm for computing a transversal on the cylinder to a set of horizontal segments.

Thus we can state the main result of this subsection

THEOREM 6.1 *Let S be a set of segments in the cylinder, then:*

a) *If S has at most one horizontal segment* PROBLEM 1 *always has a solution and a solution can be found in linear time.*

b) *If S has two horizontal segments* PROBLEM 1 *has a solution if and only if those segments are not on the same parallel, and a solution, if it exists, can be found in linear time.*

c) *If S has $N \ge 3$ horizontal segments, deciding whether or not* PROBLEM 1 *has a solution can be solved in $O(Nl)$, where l is the least* *common multiple of the denominators of the coordinates of the parallel containing the horizontal segments.*

Proof: Parts *a)* and *b)* are Lemma 6.2. If in the collection of segments S there exist horizontal and non-horizontal segments, of course the first step will be to consider the subset $S' \subseteq S$ of horizontal segments in it. Now, it is important to note that if for a given $m \in \mathbf{Z}$ the straight line joining $p = (0,0)$ with the point $(a_N + m, 1)$ is a transversal, then all straight lines joining $p = (0,0)$ with $(a_N + m + kl, 1)$ are transversals as well, where l is the least common multiple of the denominators of the coordinates of the parallels containing the horizontal segments.   □

2.2   STABBING SEGMENTS

As we have already defined in the Introduction of this Section, given a set S of N segments on the cylinder we say that a bisector b *stabs* S if each segment in S has an extreme in each part of the cylinder defined by b. Thus a problem equivalent to PROBLEM 1 arises:

PROBLEM 2: Find a bisector stabbing S.

We think that this problem is much more difficult than PROBLEM 1. In fact, we have not been able to give a polynomial algorithm for solving this problem. Thus we propose a variant in the following way. As we want to split each segment in S into two parts we label the extremes of each segment with two colors, say red and blue. Then a new, and easier, version of PROBLEM 2 can be stated:

PROBLEM 2'.- Construct a bisector that separates red points from blue points.

If we consider PROBLEM 2' in the plane instead of on the cylinder, using Megiddo's linear programming [Megiddo, 1983] a linear algorithm can be found in the following way: we try to find a straight line with equation $r \equiv \{Ax + By + C = 0\}$ such that all blue points are in one of the two half-planes that r defines and all red points are in the other; this can be translated into a system of $2N$ inequations that can be solved in linear time (obviously, this result is true in the case of the sphere). If we try to adapt this method directly to the cylinder we find an important obstacle, this obstacle being that bisectors on the cylinder are formed by two segments (instead of only one line, as in the plane), let b_1 and b_2

be those two segments. In this way each segment in S could be stabbed either by b_1 or by b_2. Therefore a naive approximation to PROBLEM 2' could be to try to check whether all segments in S are stabbed by b_1, or to check if the first segment is stabbed by b_2 and all others by b_1, and so on. Obviously this method leads to an exponential algorithm. Now we will see that, in spite of these first considerations a polynomial algorithm can be given for solving PROBLEM 2'.

THEOREM 6.2 PROBLEM 2' *can be solved in* $O(N^3)$ *time.*

Proof: Let $S = \{s_1, s_2, \ldots, s_N\}$ be a set of segments on the cylinder. The blue extreme of s_i will be denoted by B_i and its red extreme by R_i. We are looking for two points on the cylinder P_1 and P_2 such that the bisector which they define, let $b(P_1, P_2)$ denote that bisector, separates blue points from red points. In order to impose on $b(P_1, P_2)$ the condition given by an extreme of a segment, we have to know which of the two equations of the bisector we must use. This equation will be the equation of the segment that has empty intersection with the generatrix opposite to the extreme.

Let g'_{B_i} and g'_{R_i} denote the two generatrices opposite to B_i and R_i, respectively. Those generatrices divide the cylinder into $2N$ vertical strips. If we now decide that P_1 and P_2 are in given strips, it is easy to see that we can know which of the equations of $B(P_1, P_2)$ must be used for each extreme point. Thus, regarding the $2N$ vertical strips, we can choose each pair of points in a quadratic number of possible situations, and each situation can be solved in linear time (using Megiddo's algorithm). So combining these two facts, we get our result in cubic time. □

Of course, it is not clear whether the time obtained in the last theorem is optimal.

3.   VISIBILITY IN THE PRESENCE OF OBSTACLES

We will now prove that the results obtained in the last section will allow us to solve another problem related to visibility on the cylinder. More concretely, as the title of this section suggests, we are concerned with visibility in the presence of obstacles. We think that this problem has some potentially important applications in robotics, more concretely in the design of trajectories [Schwartz et al., 1987, Schwartz and Yap, 1987].

Suppose that two points are given in the cylinder and that some obstacles (segments) appear between them, a natural question is whether each point can *see* each other avoiding the obstacles, in other words, we are trying to find a geodesic joining both points and such that the intersection of this helix with the segments is the empty set. If such a geodesic exists the following question will be to find the shortest geodesic among all verifying that condition. In fact, a first question is whether there exist or not infinitely many geodesics joining both points. Of course, this is a kind of problems which cannot be formulated in the plane and which never appears in classical Computational Geometry. It is easy to see a first and partial answer to this problem.

LEMMA 6.5 *Let p and q be two points in the cylinder and let S be a set of segments in the cylinder. If there exist infinitely many geodesics joining p and q with empty intersection with S, then all the segments of S which have non-empty intersection with the horizontal band defined by p and q are horizontal.*

Proof: Suppose that in the horizontal band defined by $p = (0,0)$ and $q = (a, b)$ there exists a non-horizontal segment st (suppose that the ordinate of s is greater than the ordinate of t). Then consider the line l joining a copy of s in a tile and a copy of t in the following tile. Call C_p the intersection of l with the horizontal line containing p ($y = 0$) and C_q the intersection of l with the horizontal line containing q ($y = b$). Now, a copy of p to the right of C_p cannot see to a copy of q to the left of C_q (this is to say, the line joining both copies intersect the copy of one of the segments of S). Hence there cannot exist infinitely many geodesics joining p and q with empty intersection with S (see Figure 6.14).

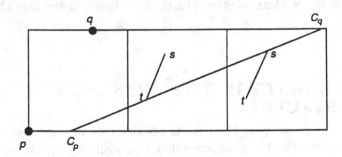

Figure 6.14.   The construction given in the proof of Lemma 6.5.

□

However, the condition given in Lemma 6.5 is not complete since in Figure 6.15 a configuration is shown with only horizontal segments between p and q and such that there exists no geodesic joining both points and avoiding the segments (all possible geodesics intersect some of the horizontal segments).

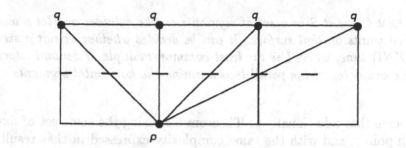

Figure 6.15. The point p does not see q, and between them there exist only horizontal segments.

At this point it is very easy to establish the connection between this question and the problem of the transversal treated at the beginning of this chapter. In fact, in Section 2 we have calculated all intersections between the geodesics joining two points and a horizontal line (see Lemma 6.3). Thus given the parallel $y = c$ we define the set of forbidden points for that parallel as the set

$$\mathcal{F}_c = \left\{ \left(\frac{c(a+m)}{b}, c \right) : m \in \mathbf{Z} \right\}.$$

Although this is an infinite set of points in $\mathcal{T}$, we have seen in Lemma 6.3 that $\mathcal{F}$ is a finite set in $\mathcal{P}$. Thus with the same considerations of Section 2 we have:

LEMMA 6.6 *Let S be a set of segments on the cylinder and let p and q be two points on that surface. If there exist infinitely many geodesics joining p and q with empty intersection with S, then all segments of S which intersect the band defined by p and q are horizontal and those segments included in the parallel $y = c$ do not cover the set $\mathcal{F}_c$. Moreover, if all segments of S which intersect the band defined by p and q are horizontal and there exists a geodesic joining p and q with empty intersection with S, then there exist infinitely many geodesics in those conditions.*

Using this lemma we can determine whether or not the set of geodesics joining two points is infinite. To obtain this goal it is enough to label the points of $\mathcal{F}_c$ as in Section 2 and to modify conveniently the algorithm TRANSV-HORIZONTAL. Of course, that algorithm will tell us not only whether the set of geodesics joining two points in the presence of obstacles is infinite, but we can extract some additional information about what is the shortest geodesic joining p and q. Finally, we can summarize the results of this section in the following:

THEOREM 6.3 *Let S be a set of segments on the cylinder and let p and q be two points on that surface. It can be decided whether or not p sees q in $O(Nl)$ time, where l is the least common multiple of denominators of the coordinates of the parallels containing the horizontal segments of S.*

Observe that additionally to Theorem 6.3, using the same set of forbidden points, and with the same complexity expressed in that result, other questions can be decided such as:

- The length of the minimum geodesic joining two points p and q in the cylinder.

- How many times can a point see another point.

- Whether a point can see another point infinitely many times.

4.   NOTES AND COMMENTS

In this chapter we have visited one of the classical problem of Computational Geometry such as that of stabbing a segment set. In contrast to what happened with the width of a convex set (see Chapter 4), here the two alternative definitions do not lead to a common solution, and so we have considered two different problems: to find transversal helices, and stabbing segments with bisectors. Each one of those cases leads to interesting considerations, especially from the complexity of the algorithms we have obtained. Nevertheless, we do not know whether the algorithms which we present here are optimal algorithms or not.

In spite of their naturalness, we are not aware of any previous work by other authors considering these transversal problems. The same does not happen with visibility questions such as those considered in Section 3., and we can mention circular visibility first introduced in [Agarwal and Sharir, 1993]. The finiteness of the visibility multigraph on the cylinder and on the torus have been considered in [Cobos et al., 1995] and some art-gallery type problems on the cylinder were studied in [Dana et al.,

1995]. In addition, some of the results of this chapter have appeared previously in [Dana et al., 1998].

Chapter 7

TRIANGULATIONS

With convex hulls and Voronoi diagrams, one of the most studied problems in Computational Geometry is that of constructing a triangulation of a polygon or of a set of sites. A triangulation is a partition of the domain defined by the input into triangles which meet only at shared sides. Since this kind of meshes are needed in all domains where the ambient space must be discretized, this structure must be studied on surfaces in addition to the plane.

1.   INTRODUCTION

One of the most useful tools emanating from Computational Geometry is the triangulation of a polygon or of a set of points, that is, a partition of the domain defined by the input (and we will see later that an interesting problem in our surfaces is to recognize what that domain is) into triangles which meet only at shared sides. In the plane that definition agrees with a maximal planar subdivision, using straight line segments between vertices in the set.

The applications of triangulations range from the finite-element method [Strang and Fix, 1973, Cavendish, 1974] to numerical interpolation [Preparata and Shamos, 1985]. Solid modeling or robotics are other areas that have used triangulations as an important tool [de Berg et al., 1997] [Goodman and O'Rourke, 1997].

Finally, in the context of Computational Geometry the triangulation of a set of sites or of a polygon is a very useful tool in its own right for its use in other problems, for instance in some point location algorithms, in polygon intersection algorithms, and in some visibility problems.

127

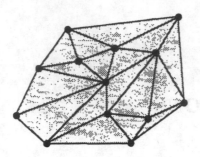

Figure 7.1.   A triangulation of a set of points in the plane.

Of course, all problems and application mentioned above can appear naturally in the case of our surfaces. Thus, it is natural that this book, which is dedicated to 'translating' classical Computational Geometry problems to surfaces, must contain a chapter devoted to triangulations.

As in other structures studied before, some important differences appear in this subject. For instance, we can give two possible definitions of triangulations which are equivalent in the planar case (a maximal subdivision and a triangular subdivision), and it is not clear whether both definitions agree outside the plane (we will see that both are equivalent in the case of the cylinder or the sphere but not in the case of the torus). Another important problem studied in this chapter is: what is the domain defined by a set of sites when we perform a triangulation? This problem is presented mainly in the case of the cylinder, although we will mention in two sections what happens with the sphere and the torus for both set of sites and polygons.

The two last problems considered here refer to the graph of triangulations, so we try to solve the questions: Is it possible to go from a triangulation to another using diagonal flips? and, how can we obtain optimal triangulations? We will settle the former in the positive for the surfaces considered here but in the negative for general surfaces. And we will see that the usual planar strategies are not valid in surfaces, so some other methods must be used.

2.   TRIANGULATIONS ON THE CYLINDER

In order to generalize the definition of a triangulation of a set of points in the plane to the cylinder, we will substitute on this surface, as we have done with other structures, straight line segments by geodesic segments (segments, for short).

So given a set P of N points on the cylinder a *pseudo-triangulation of P* is a maximal set of segments connecting points of the set, that is, no segment can be added without intersecting one of the existing segments. One of the first questions we will face is whether a pseudo-triangulation is actually a triangulation; i.e., a partition of the domain defined by the set P into triangles which meet only at shared sides. Obviously, from the definition of triangulation, the other initial question arises of 'what is the domain defined by P?'.

We will consider only sets in non-Euclidean position on the cylinder, since, otherwise, computing triangulations (or pseudo-triangulations) is similar to the planar case, because all segments joining points pairwise will be contained between the two opposites generatrices which enclose the set, see Figure 7.2.

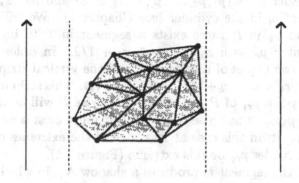

Figure 7.2.   A pseudo-triangulation of a set of points in Euclidean position on the cylinder agrees with the triangulation of the planar projection of the points in the set.

So let $P = \{p_1, p_2, \ldots, p_N\}$ be a set of N $(N > 3)$ points on the cylinder in non-Euclidean position, and let T be a pseudo-triangulation of it. The first difference that we find with the case of the plane is that while every triangulation of a set of N points in the plane defines only one unbounded face, in the case of the cylinder we will prove that every pseudo-triangulation of P defines in this surface two unbounded faces.

Figure 7.3. A pseudo-triangulation of a set of points in non Euclidean position on the cylinder.

LEMMA 7.1 *Any point of P is the extreme of at least one segment of T to its left and one segment of T to its right.*

Proof: Consider $P = \{p_1, p_2, \ldots, p_N\}$ $(N > 3)$ and let (x_i, y_i) be the coordinates of p_i in the cylinder (see Chapter 1). We will prove that for any point p_k in P there exists a segment in T to its right, that is, a segment $p_k p_r$ with $x_k \leq x_r \leq x_k + 1/2$. In order to do that we consider two copies of the cylinder and the vertical strip defined by $S_k = \{(x,y); x_k < x < x_k + 1/2\}$, see Figure 7.4. This strip must contain at least one point p_{r_1} of P $(p_{r_1} \neq p_k)$, otherwise P will be in Euclidean position. If $p_k p_{r_1}$ is not a segment of T it must exist a segment t_2 in T intersecting it; in this case at least one of the extremes of t_2 will be contained in S_k, let p_{r_2} be this extreme (Figure 7.4).

Note that the segment t_2 produces a shadow $\mathcal{N}_1$, in which p_{r_1} is contained, in such a way that there is no segment in T joining p_k to a point in $\mathcal{N}_1$, (Figure 7.5).

Suppose that $p_k p_{r_2}$ is not a segment in T, then there exists a segment t_3 in the pseudo-triangulation without extremes in $\mathcal{N}_1$ such that it has, at least, an extreme in S_k. This segment produces a new shadow $\mathcal{N}_2$, in which p_{r_2} is contained, and such that there is no segment in T joining p_k to a point in $\mathcal{N}_2$, (Figure 7.6).

If we repeat this process, since P has a finite number of elements we will find a point p_{r_n} such that $p_k p_{r_n} \in T$, and the result holds.   □

If we call *essential polygon* of P any closed polygonal curve joining points of the set P which wraps around the cylinder (in other words, it is not homotopically trivial), it is possible to define a partial ordering in

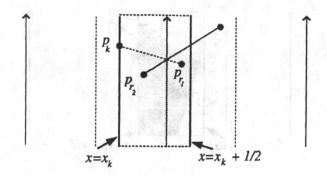

Figure 7.4. The shadow strip is the set $S_k = \{(x, y); x_k < x < x_k + 1/2\}$ (see proof of Lemma 7.1).

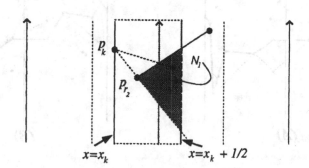

Figure 7.5. There is no segment in T joining p_k with a point in $\mathcal{N}_1$ (see proof of Lemma 7.1).

the set of essential polygons of P as follows: given two essential polygons $\mathcal{P}_1$ and $\mathcal{P}_2$ we will say that $\mathcal{P}_1 < \mathcal{P}_2$ if the intersection of $\mathcal{P}_1$ and every generatrix has smaller ordinate that the intersection of this generatrix with $\mathcal{P}_2$ (see Figure 7.7). An *upper polygon* (resp., *lower polygon*) is a maximal essential polygon (resp., minimal essential polygon) with this ordering.

THEOREM 7.1 *Any pseudo-triangulation of a set of N points in non-Euclidean position on the cylinder defines on this surface two unbounded faces delimited by an upper and a lower polygon respectively.*

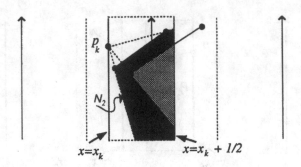

Figure 7.6. There is no segment in T joining p_k with a point in $\mathcal{N}_2$ (see proof of Lemma 7.1).

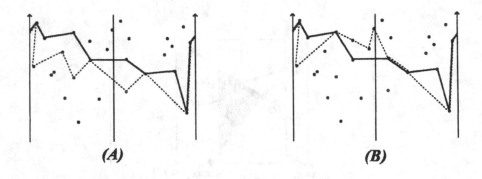

(A) *(B)*

Figure 7.7. Let $\mathcal{P}_1$ be the dark essential polygon in the picture and let $\mathcal{P}_2$ be the discontinuous one:*(A)* $\mathcal{P}_2 < \mathcal{P}_1$; *(B)* $\mathcal{P}_1$ and $\mathcal{P}_2$ are not related.

Proof: Taking into account Lemma 7.1 we know that every point of the set $P = \{p_1, p_2, \ldots, p_N\}$ is the left extreme of at least one segment of the pseudo-triangulation T and the right extreme of another. Starting, for instance, in p_1 we can 'walk' on a segment to its right and arrive at another point p_k, and continue 'walking' on a segment to the right of p_k until arriving at a new point p_j, see Figure 7.9.

As we have only a finite set of points, eventually the polygonal that we are walking on must find a point considered before. So we have found a closed polygonal curve wrapping around the cylinder (or essential polygon), and therefore it defines in this surface two unbounded faces.

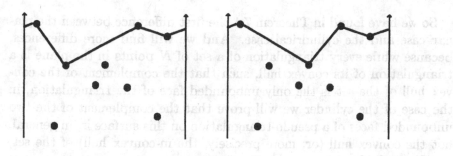

Figure 7.8. For the same set of points we can have more than one upper polygon.

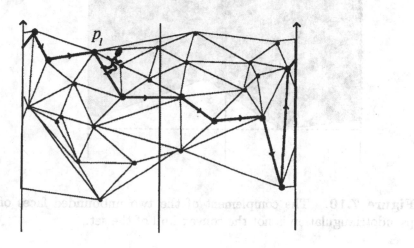

Figure 7.9. If we 'walk' from p_1 always to the right we will describe an essential polygon (see proof of Theorem 7.1).

It remains to prove that the border of these two unbounded faces are, respectively, an upper and a lower polygon. But, if the border of the upper unbounded face is not maximal with the partial ordering defined before, it must exist another essential polygon greater than the essential polygon on the border, so, it will be possible to add new edges to the pseudo-triangulation, which contradicts the maximality of T. Analogously with the lower unbounded face.   □

So we have found in Theorem 7.1 the first difference between the planar case and the cylindrical case. And we will find more differences, because while every triangulation of a set of N points in the plane is a triangulation of its convex hull, such that the complement of the convex hull of the set is the only unbounded face of the triangulation, in the case of the cylinder we will prove that the complement of the two unbounded faces of a pseudo-triangulation on this surface is, in general, not the convex hull (or, more precisely, the m-convex hull) of the set, see Figure 7.10.

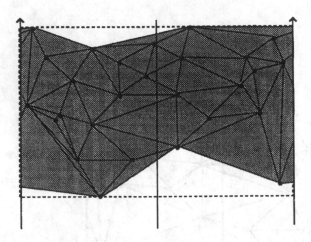

Figure 7.10. The complement of the two unbounded faces of this pseudotriangulation is not the convex hull of the set.

But we will prove that, as in the plane, each bounded region of the pseudo-triangulation is a triangle, that is, that every pseudo-triangulation on the cylinder is actually a triangulation. In order to obtain that result we need to introduce a new concept: we call *Euclidean polygon* any bounded region delimited by a single closed polygonal curve defined by segments on the surface, the vertices of which are said to be *vertices* of the Euclidean polygon. Observe that as the region delimited by the closed polygonal is bounded then this polygonal cannot be essential, and the region must be homotopically trivial (i.e., simply connected), see Figure 7.11.

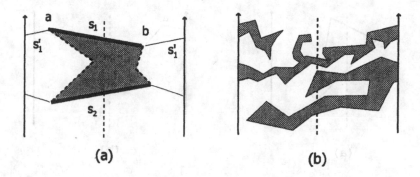

Figure 7.11. (a) The bounded region in this picture is not an Euclidean polygon because s_1 and s_2 are not segments on the cylinder, the segment defined by a and b is s_1'; (b) An Euclidean polygon on the cylinder.

A vertex of an Euclidean polygon is called *convex* if its internal angle is strictly less than π.

LEMMA 7.2 *Every Euclidean polygon on the cylinder must have at least one convex vertex.*

Proof: As in the case of polygons in the plane [O'Rourke, 1994], all vertices having minimum or maximum ordinate, except those between two vertices with the same ordinate, are convex vertices. □

A *diagonal* of an Euclidean polygon is a segment between two of its vertices which is internal to the polygon (all the vertices of the diagonal must lie on the interior of the polygon).

LEMMA 7.3 *Every Euclidean polygon on the cylinder with four or more vertices has a diagonal.*

Proof: The proof is based on the proof of Meister's Lemma [Meister, 1975] given by O'Rourke in his book [O'Rourke, 1994]. Consider an Euclidean polygon $\mathcal{P}$; let v be a convex vertex of it having minimum ordinate (Lemma 7.2); and let a and b the vertices adjacent to v. Note that we have two possibilities: (a) $\{v, a, b\}$ are in Euclidean position on the cylinder, then avb is a bounded triangle on the cylinder ; or (b) $\{v, a, b\}$ are in non-Euclidean position and, in this case, avb is an essential polygon.

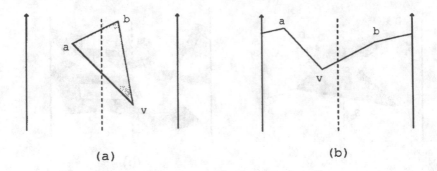

(a)   (b)

Figure 7.12. (a) If $\{v, a, b\}$ are in Euclidean position in the cylinder, avb is a bounded triangle on the cylinder; (b) if $\{v, a, b\}$ are in non-Euclidean position, avb is an essential polygon (see proof of Lemma 7.3).

In the first case, when avb is an Euclidean triangle, we will proceed as in [O'Rourke, 1994]: if ab is a diagonal, we have finished; otherwise, either ab is exterior to $\mathcal{P}$, or it intersects $\partial\mathcal{P}$. In any case, since $\mathcal{P}$ has more than three vertices there must exist at least one vertex into the closed triangle $\triangle avb$; let x be the nearest one to v, where the distance is measured orthogonally to the circle ab, then vx must be a diagonal.

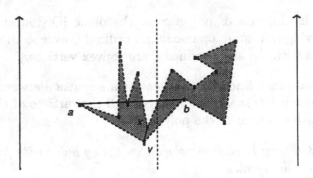

Figure 7.13. If x is the nearest vertex to v in the closed triangle $\triangle avb$, the segment vx is a diagonal of $\mathcal{P}$ (see proof of Lemma 7.3).

In the last case, when $\{v, a, b\}$ are in non-Euclidean position and avb is an essential polygon, we consider the rays from v through a and b, respectively, and we 'sweep' rotating these rays centered in v toward the generatrix containing v as Figure 7.14 shows.

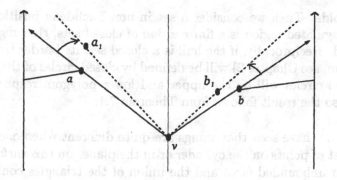

Figure 7.14. Sweeping with the rays va and vb toward the generatrix containing v (see proof of Lemma 7.3).

Let a_1 and b_1 be the first vertices of $\mathcal{P}$ found by va and vb, respectively, sweeping as we have seen before. We test whether $a_1 v b_1$ is a bounded triangle or an essential polygon. In the first case, we proceed as in O'Rourke's book, otherwise we continue the sweep in the same way and we will find a_2 and b_2 and we will ask again about the Euclidean position or not of $a_2 v b_2$. Repeating this process, and since $\mathcal{P}$ is a polygon, we must find a_k and b_k such that $a_k v b_k$ is an Euclidean triangle, and then we will have the result.   □

As an immediate consequence of Lemma 7.3 we obtain two results which in this case coincide with the planar case. The first of those results establishes that any pseudo-triangulation on the cylinder is in fact a triangulation, and the second enunciates the same result for Euclidean polygons.

COROLLARY 7.1 *Each bounded region of any pseudo-triangulation of a set of points in the cylinder is an Euclidean triangle.*

COROLLARY 7.2 *Every Euclidean polygon in the cylinder admits a triangulation.*

However, the next consequence of Theorem 7.1 gives another difference with the planar case.

COROLLARY 7.3 *A triangulation of a set of points P in the cylinder is a triangulation of its m-convex hull if and only if its hull is a closed set.*

Proof: As we know, if the set of points is in Euclidean position in the cylinder, its convex hull is similar to the plane, and in the plane this result holds. Then we consider a set in non-Euclidean position. Since any triangulated region is a finite union of closed sets, that region must be closed. Reciprocally, if the hull is a closed set, its border (m-top and m-bottom, see Chapter 2) will be defined by closed circles of the cylinder, and these circles will be the upper and lower polygon, respectively, of the set; so the result follows from Theorem 7.1.   □

So far we have seen that things are quite different when one triangulates a set of points on the cylinder or in the plane: on this surface, there exist two unbounded faces and the union of the triangles could not be the convex hull of the set, and, moreover, as we can see in Figure 7.15, this triangulated region might not be unique.

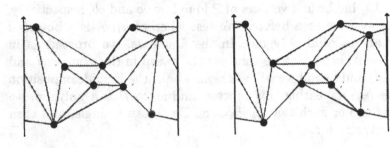

Figure 7.15.   For the same set of points, the triangulated region might not be unique

The natural reaction is now to ask if the triangulated region is unique in some case and if it is possible to determine it by looking at the set of points. We start testing whether a point of the set belongs to an upper or lower polygon. Given $P = \{p_1, p_2, \ldots, p_N\}$ on the cylinder, let (x_i, y_i) be the coordinates of p_i. Let $g_k^+ = \{(x_k, y); y \geq y_k\}$; we call L_k the first point in $\{(x, y); x_k - 1/2 < x < x_k\}$ which is found when we sweep g_k^+ rotating counter-clockwise and centered in p_k; and let R_k be the first point found in $\{(x, y); x_k < x < x_k + 1/2\}$ sweeping g_k^+ clockwise.

LEMMA 7.4 *The point p_k belongs to an upper polygon if and only if g_k^+ does not intersect the segment $L_k R_k$.*

Proof: We see at once that if p_k belongs to an upper polygon, its neighbors, the adjacent vertices in the polygon, must be L_k and R_k.

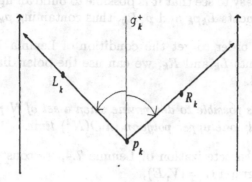

Figure 7.16. L_k is the first point in $\{(x,y); x_k - 1/2 < x < x_k\}$ that is found rotating g_k^+ counter-clockwise; and R_k is the first one in $\{(x,y); x_k < x < x_k + 1/2\}$ sweeping g_k^+ clockwise.

Suppose that p_k is in an upper polygon $\mathcal{P}$, if g_k^+ intersects the segment $L_k R_k$, it is possible to add this segment to the triangulation which contradicts the maximality of it as a maximal set of segments.

Conversely, if we have that p_k does not intersect the segment $L_k R_k$, it might happen that: a)$\{L_k, p_k, R_k\}$ are in Euclidean position and $L_k R_k$ intersects $g_k^- = \{(x_k, y); y < y_k\}$; or b) $\{L_k, p_k, R_k\}$ are in non-Euclidean position, $L_k p_k R_k$ divides the cylinder into two unbounded faces, and $L_k R_k$ does not intersect $g_k^- = \{(x_k, y); y < y_k\}$, see Figure 7.17.

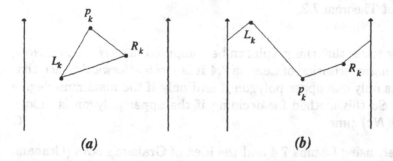

(a)   *(b)*

Figure 7.17. In both situations it is possible to build an upper polygonal containing the segments $L_k p_k$ and $p_k R_k$, so containing p_k.

In any case it is easy to see that it is possible to build an upper polygon containing the segments $L_k p_k$ and $p_k R_k$, thus containing p_k.   □

Observe that in order to get the condition of Lemma 7.4, or more specifically the points L_k and R_k, we can use the polar diagram introduced in Chapter 4.

THEOREM 7.2 *It is possible to determine when a set of N points on the cylinder defines only one upper polygon in $O(N^2)$ time.*

Proof: Using the characterization of Lemma 7.4, we construct the following undirected graph $G = (V, E)$,

$$V = \{p_k \; ; \; g_k^+ \text{ does not intersect to } L_k R_k\},$$

$$E = \{(p_i, p_j) \; ; \; p_j = L_i \text{ or } p_j = R_i\},$$

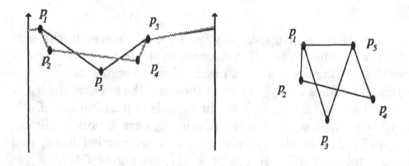

Figure 7.18. Two upper polygons and the associated graph given in the proof of Theorem 7.2.

It is easy to see that this graph can be computed in $O(N^2)$ time. Now, using the characterization of Lemma 7.4 it is straightforward to see that there exists only one upper polygon if and only if the maximum degree of G is 2. So this method for deciding if the upper polygon is unique requires $O(N^2)$ time.   □

Moreover, using Lemma 7.4 and the idea of Graham's scan [Graham, 1972] for computing planar convex hulls, we are now able to design the following algorithm for computing an upper polygon for a set of points, $P = \{p_1 = (x_1, y_1), p_2 = (x_2, y_2), \ldots, p_N = (x_N, y_N)\}$, on the cylinder: consider $y_M = \max\{y_1, y_2, \ldots, y_N\}$ and sort circularly the points by their abscissae, starting from one of them with ordinate y_M, this point is labeled as START. Repeatedly examine triples of consecutive points (p_{i-1}, p_i, p_{i+1}):

1. If the segment $p_{i-1}p_{i+1}$ intersects g_i^+, eliminate vertex p_i and consider $(p_{i-1}, p_{i+1}, p_{i+2})$;

2. Otherwise consider (p_i, p_{i+1}, p_{i+2})

The scan terminates when it advances all the way around to reach START again, see Figure 7.19.

Note that the role played in that algorithm by Graham's scan, can be played as well by the polar diagram presented in Chapter 4. Thus, we can enunciate:

THEOREM 7.3 *Given a set of N sites P on the cylinder, it is possible to construct one of the upper polygons of P in optimal time $O(N \log N)$.*

The complexity expressed in Theorem 7.3 follows from the complexity time of Graham's algorithm [Graham, 1972].

Summarizing, and from a theoretical point of view, we have seen that every triangulation of a set of points in non-Euclidean position on the cylinder defines two unbounded faces, the complement of which will be the convex hull of the set if and only if the hull is closed. We have also seen that every bounded face is a triangle. About the unbounded faces, we have shown that their borders are an upper and a lower polygon and that they might be not unique, so we can triangulate different regions each time. From an algorithmic point of view we have seen that it is possible to know when the upper polygon is unique in $O(N^2)$ time, and to construct a upper polygon in optimal $O(N \log N)$ time.

2.1   MAXIMIZING THE SMALLEST ANGLE

There exist many criteria for what constitutes a 'good' triangulation for several purposes, some of which involve maximizing the smallest angle [George, 1981, Field, 1986].   In this way, for the sake of many applications of interpolating on a surface, it is more desirable to have 'fat' triangles (nearly equilateral), both for the purposes of a graphical display and for calculations in numerical analysis. So since the ability to perform good triangulations seems to be an useful tool, we will try to do so with triangulations of set of points on the cylinder. In order to do this we define the following order: for each triangulation T with k triangles we index their internal angles $(\alpha_1, \alpha_2, \ldots, \alpha_{3k})$ in such a way that $\alpha_i \leq \alpha_j$

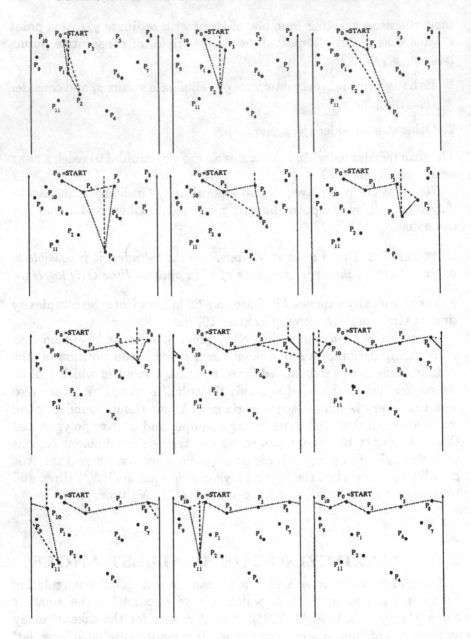

Figure 7.19. Twelve steps of the construction of an upper polygon, see text for details.

for $i < j$, and we define the *vector angle of* T as $A(T) = (\alpha_1, \alpha_2, \ldots, \alpha_{3k})$. Given two triangulations T and T', with $A(T') = (\alpha'_1, \alpha'_2, \ldots, \alpha'_{3m})$ we say that $A(T) \succ A(T')$ when $A(T)$ is lexicographically greater than $A(T')$. We say that a triangulation T of a set of points *maximizes the minimum angle* when $A(T) \succeq A(T')$ for any other triangulation T' of the set, equivalently T is said to be *equiangular*. For a set of sites in the plane the Delaunay triangulation is maximal with respect to the ordering defined above, and this triangulation can be obtained as the dual graph of the Voronoi diagram of the set. Since we know how to construct the Voronoi diagram of a set of sites on the cylinder (see Section 4.2) one can try to do this in the cylinder, namely, to construct the Delaunay triangulation from the Voronoi diagram as its dual. But unfortunately, as we can see in Figure 7.20, the dual of the Voronoi diagram of a set of points in the cylinder can not be a triangulation.

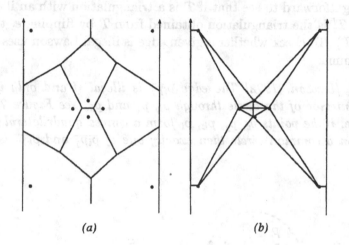

(a)   (b)

Figure 7.20. (a) The Voronoi diagram of a set of eight points on the cylinder; (b) The dual of the Voronoi diagram in (a), it is not a triangulation on the cylinder.

Even if we eliminate overlapping edges, obtaining the Euclidean dual, we do not obtain a triangulation.

So it is necessary to look for another method. We can try to adapt to the cylinder the idea of [Lawson, 1972] based on transforming the initial triangulation by a sequence of *flips*. Given an edge $e = \overline{p_i p_j}$ of a triangulation of a set of sites in the plane, if it is not an edge of the unbounded face it is incident upon two triangles $p_i p_j p_k$ and $p_i p_j p_l$. If $p_i p_j p_k p_l$ is a convex quadrilateral we can replace $\overline{p_i p_j}$ by the other diagonal of this quadrilateral, and we call this operation an *edge flip*. An edge e is said to be *illegal* if it is possible to increase locally the smallest angle by flipping that edge.

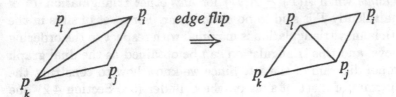

Figure 7.21.   Flipping an edge

It is straightforward to see that if T is a triangulation with an illegal edge e, and T' is the triangulation obtained from T by flipping e, then $A(T') \succ A(T)$. To check whether a given edge is illegal Lawson uses the following lemma:

LEMMA 7.5 *[Lawson, 1972] The edge $\overline{p_i p_j}$ is illegal if and only if p_l lies in the interior of the circle through p_i, p_j and p_k, see Figure 7.22. Furthermore, if the points p_i, p_j, p_k, p_l form a convex quadrilateral and do not lie on a common circle then exactly one of $\overline{p_i p_j}$ and $\overline{p_k p_l}$ is an illegal edge.*

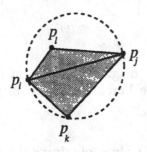

Figure 7.22.   Checking whether an edge is illegal in a planar triangulation.

So, given an initial triangulation in the plane it is possible to reach another triangulation which maximizes the minimum angle just flipping edges until all edges are legal. Unfortunately, we will find again that it is not possible to adapt this method to our surface. Given an edge $e = \overline{p_i p_j}$ of a triangulation on the cylinder, if it is not an edge of the unbounded faces it is incident upon two triangles $p_i p_j p_k$ and $p_i p_j p_l$. If $p_i p_j p_k p_l$ is a convex quadrilateral we can flip $\overline{p_i p_j}$ by the other diagonal of this quadrilateral if and only if this other diagonal is inside the convex quadrilateral $p_i p_j p_k p_l$, see Figure 7.23.

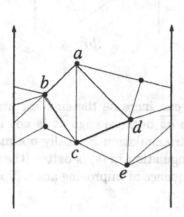

Figure 7.23. It is not possible to flip $\overline{ac}$ by $\overline{bd}$ because the latter is exterior to the quadrilateral $abcd$, but it is possible to flip $\overline{cd}$ by $\overline{ae}$.

And, as happens in the plane, an edge e is said to be *illegal* if it is possible to increase locally the smallest angle by flipping that edge. As we have said before and as we can see in Figure 7.24, in the case of the cylinder it will be not possible to reach a triangulation with optimal angles flipping edges until all edges are legal.

Searching in the literature for another method, we found in Brown's PhD Thesis [Brown, 1980] that, on the sphere, he proceeded as follows. Given the S a set of $N > 3$ sites on the sphere such that no four points are co-circular (no four points lie on a same circle on the sphere), construct the 3D convex hull of the sites; it is not difficult to see that the edges and facets of this convex hull give the Delaunay triangulation on the sphere, which maximizes the minimum angle. But again we cannot adapt this idea to our surface, as we can see in [Cortés et al., 1998], since the result of projecting the 3D convex hull of the set of sites on the cylinder is not, in general, a triangulation, or even if a triangulation is obtained, this triangulation could be no optimal, as Figure 7.25 shows.

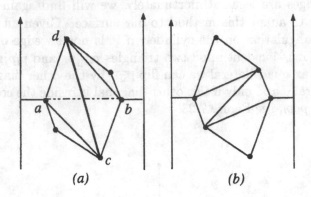

(a)   (b)

Figure 7.24. (a) No flip can increase the smallest angle, (it is not possible to flip edge $\overline{cd}$ with $\overline{ab}$ because edge $\overline{ab}$ is not interior to the quadrilateral $abcd$), so this triangulation is locally optimal, that is, all edges are legal; but, the triangulation in (b) is better (the smallest angle is greater) and there is no sequence of improving angle flips which allows to obtain it from (a).

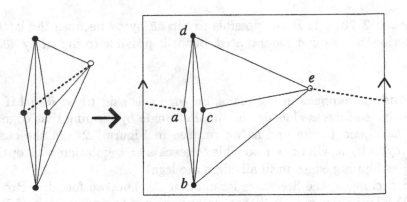

Figure 7.25. The projection of the 3D convex hull on the cylinder (orthogonally with respect to its edge) leads to the triangulation in the right picture. This triangulation can be improved (maximizing its smallest angle) just by flipping bd into ac.

We have received a lot of 'bad news' when we have tried to adapt planar algorithms in order to maximize the minimum angle in a triangulation in the cylinder; but, finally, we will find an algorithm that is adaptable to cylinder. In [Guibas et al., 1992] a different approach to Delaunay triangulations is described: to compute this optimal triangulation directly using a randomized incremental algorithm. We briefly describe it: start with a large triangle $p_1p_2p_3$ which contains the set P; compute the Delaunay triangulation of $P \cup \Omega$ where $\Omega = \{p_1, p_2, p_3\}$, and later discard Ω together with all incident edges. For this purpose we have to choose p_1, p_2 and p_3, far enough away from the points of the initial set P, in particular, we must ensure they do not lie in any circle defined by three points in P.

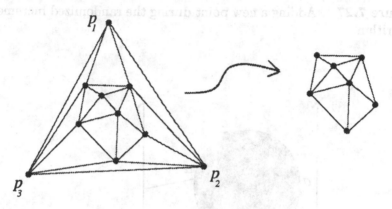

Figure 7.26. Computing a Delaunay triangulation using a randomized incremental approach.

Since the algorithm is randomized incremental, it adds the points in random order and it maintains a Delaunay triangulation of the current point set. When we add p_j we must find the triangle which contains it and we add edges from p_j to the vertices of this triangle; if p_j fall on an edge e of the triangulation add edges from p_j to the opposite vertices in the triangles sharing e, see Figure 7.27. Next we ask if the edges of the new triangles are legal or not; when an edge is illegal replace it by a legal one through edge flips.

This randomized incremental algorithm in the plane is due to Guibas et al [Guibas et al., 1992]; and a detailed description of it can be found in the book of de Berg et al. [de Berg et al., 1997].

When one tries to adapt the above algorithm to the cylinder the first difference that one finds is that on this surface it is possible to have a situation as Figure 7.28 shows.

So we must redesign the test given in the plane by Lemma 7.5 for checking whether an edge of a triangulation on the cylinder is illegal or

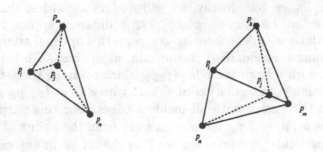

Figure 7.27. Adding a new point during the randomized incremental algorithm.

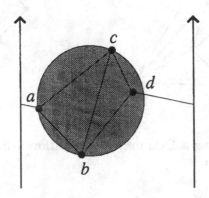

Figure 7.28. The point d lies inside the circle defined by a, b and c, but we cannot flip cd to ab, since ab is exterior to the quadrilateral defined by the two triangles sharing cd.

not. In order to do that, given three points $A_i = (x_i, y_i)$, $1 \leq i \leq 3$, with $0 \leq x_1 \leq x_2 \leq x_3 \leq 1$, we call the *Euclidean circle* associated with A_1, A_2 and A_3 the intersection of the planar circle containing these three points with the strip $\{(x,y); x_3 - 1/2 \leq x \leq x_1 + 1/2\}$.

LEMMA 7.6 *The edge $\overline{p_i p_j}$ is illegal if and only if p_l lies in the interior of the Euclidean circle through p_i, p_j and p_k, see Figure 7.30.*

Proof: The proof of this result is based on the result of Lemma 7.6. If p_l lies in the interior of the Euclidean circle through p_i, p_j and p_k, then

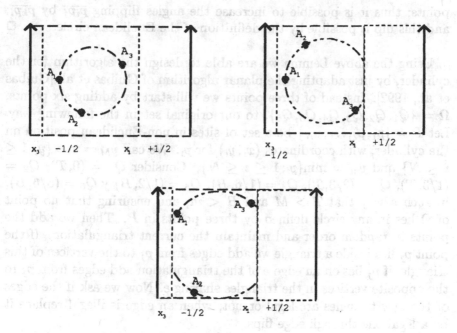

Figure 7.29. Euclidean circles associated with several sets of three points.

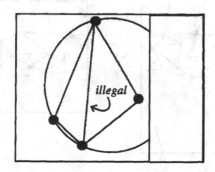

Figure 7.30. Checking whether an edge is illegal in a cylindrical triangulation.

obviously p_l lies in the interior of the planar circle through these three points; thus it is possible to increase the angles flipping $\overline{p_i p_j}$ by $\overline{p_k p_l}$; and this flip is possible by the definition of the Euclidean circle.   □

Using the above Lemma we are able to design the algorithm for the cylinder, by just adapting the planar algorithm of Guibas et al. [Guibas et al., 1992]. Instead of three points we will start by adding six points, $\Omega = \{Q_1, Q_2, Q_3, Q_4, Q_5, Q_6\}$, to our original set in the following way. Let $P = \{p_1, p_2, \ldots, p_N\}$ be a set of sites in non-Euclidean position on the cylinder, with coordinates (x_i, y_i) for p_i. We call $y_M = \max\{y_i; 1 \leq i \leq N\}$ and $y_m = \min\{y_i; 1 \leq i \leq N\}$. Consider $Q_1 = (0, T)$, $Q_2 = (1/3, T)$, $Q_3 = (2/3, T)$, $Q_4 = (1/6, B)$, $Q_5 = (1/2, B)$ y $Q_6 = (5/6, B)$, in such a way that $T > M$ and $B < m$, and ensuring that no point of Ω lies in any circle defined by three points in P. Then we add the points in random order and maintain the current triangulation. If the point p_j lies inside a triangle we add edges from p_j to the vertices of this triangle; if p_j lies on an edge e of the triangulation add edges from p_j to the opposite vertices in the triangles sharing e. Now we ask if the edges of the new triangles are legal or not; when an edge is illegal replace it by a legal one through edge flips.

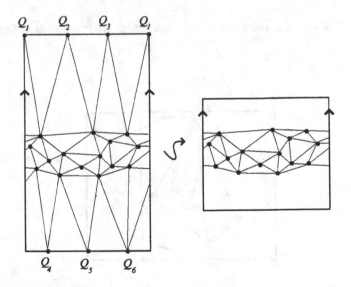

Figure 7.31.   The randomized incremental algorithm in the cylinder starts adding six new points to the original set of sites.

As the analysis of the algorithm is similar to the planar case we have:

THEOREM 7.4 *Given a set of N sites on the cylinder it is possible to compute a triangulation maximizing the smallest angle in $O(N \log N)$ expected time, using $O(N)$ expected storage.*

2.2   GRAPH OF TRIANGULATIONS

It is well known that the *graph of triangulations* of a polygon $\mathcal{P}$ in the plane is connected, this graph being $\mathcal{G}(\mathcal{P}) = (V, E)$, such that its nodes V are all possible triangulations of $\mathcal{P}$, and two triangulations are adjacent if they can be transformed into each other by one flip (Figure 7.32). A proof of this property can be found in the work of Hurtado and Noy [Hurtado and Noy, 1996].

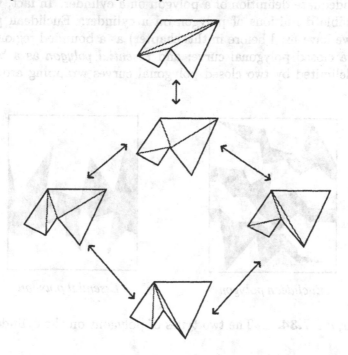

Figure 7.32.   Graph of triangulations of a polygon in the plane.

In the same way it is possible to define the graph of triangulations for a set of N sites in the plane. Since, as is well known, every triangulation of a set of sites can be transformed by flips (using, for instance, Lawson's algorithm [Lawson, 1972]) into the Delaunay triangulation of the set, the graph of triangulations of a set of sites in the plane is also connected.

$$T_1 \xrightleftharpoons[]{\text{flips}} \boxed{Delaunay\ Triangulation} \xrightleftharpoons[]{\text{flips}} T_2$$

Figure 7.33.   Given two triangulations, T and T', of a set of sites in the plane, it is possible to transform one into the other by flips, by just passing through the Delaunay triangulation of the set.

So one can think of studying whether this kind of graphs on the cylinder are connected or not, both for triangulations of a set of sites and for triangulations of polygons, and this has been done in [Cortés et al., 1999].

As in many other cases considered in this book, first of all we have to find an adequate definition of a polygon on a cylinder. In fact, we have two possible definitions of polygon on a cylinder: Euclidean polygon (which we have used before in this chapter) as a bounded region delimited by a closed polygonal curve; and *essential polygon* as a bounded region delimited by two closed polygonal curves wrapping around the cylinder.

Euclidean polygon   *Essential polygon*

Figure 7.34.   The two types of polygons on the cylinder.

THE GRAPH OF TRIANGULATIONS OF EUCLIDEAN POLYGONS ON THE CYLINDER

It is easy to respond in the affirmative to the question of the connectivity of the graph of triangulations of an Euclidean polygon on the cylinder. In fact, although not all given proofs of the same property in the plane can be adapted to the cylinder, we check the validity of some of them. So for Euclidean polygons we have:

THEOREM 7.5 *Given two triangulations of an Euclidean polygon of $N \geq$ 4 on the cylinder it is possible to transform each into the other by a sequence of flips.*

Proof: First of all, observe that any triangulation of an Euclidean polygon in the cylinder has at least two non-overlapping ears (an *ear* of a triangulation is a triangle which has only one internal diagonal), the proof of this property follows from the proof of Meister's Two Ears Theorem given in [O'Rourke, 1994], since it is easy to see that, as in the plane, the dual of the triangulation of an Euclidean polygon is a tree; then an ear is associated with a leaf node, and a tree of two or more nodes must have at least two leaves.

Consider now two triangulations T_1 and T_2 of an Euclidean polygon of $N \geq 4$ vertices $\mathcal{P}$. We proceed by induction on the number of vertices. If T_1 and T_2 have a common ear, it suffices to 'cut' by this ear and apply the induction hypothesis. Otherwise, let e_1 and e_2 be the vertices of an ear in T_1 and an ear in T_2, respectively. Without loss of generality we can assume that e_1 and e_2 are not adjacent (otherwise we can change the chosen ears, and we can find at least a couple of non-adjacent ears, one in T_1 and the other in T_2).

Now remove e_1 from $\mathcal{P}$ to obtain a polygon with $N-1$ vertices, draw in this polygon the diagonal forming an ear in e_2, and complete this diagonal to a triangulation of the $N-1$ polygon; add e_1 and we have a new triangulation $\widehat{T}_1$ of $\mathcal{P}$, with an ear common to T_1 and another one common to T_2, see Figure 7.35. Now, by just 'cutting' ears and using the induction hypothesis we can go from T_1 to T_2 by flips, through $\widehat{T}_1$, $T_1 \hookrightarrow \widehat{T}_1 \hookrightarrow T_2$.

$\square$

THE GRAPH OF TRIANGULATIONS OF A SET OF SITES ON THE CYLINDER

The next step now must be to study the graph of triangulations of essential polygons on the cylinder. But before dealing with essential polygons we will study the graph of triangulations of sets of sites, and we will have the result for essential polygons as a particular case.

So let $P = \{p_1, p_2, \ldots, p_N\}$ be a set of sites in non-Euclidean position on the cylinder (if P is in Euclidean position its graph of triangulations

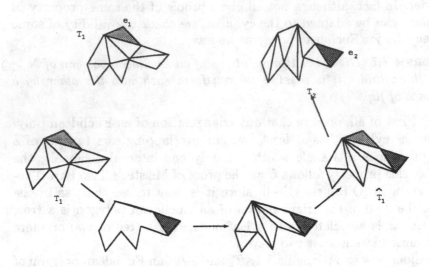

Figure 7.35. Any two triangulations of an Euclidean polygon can be transformed into each other by a sequence offlips.

is trivially connected, since this case is similar to the plane). We will have easily that the graph of triangulations of P could be non-connected. In fact, since we know that for the same set of sites it is possible to get two triangulations with different borders, and no flip changes edges in the unbounded face. We conclude that in the case of the cylinder the graph of triangulations of a set of sites might be non-connected.

Although, as we have seen, in general the answer to the question about the connectivity of the graph of triangulations of a set of sites in the cylinder is negative, we have seen that the problems occur when two triangulations do not have the same border. Thus what about two triangulations of a set of sites sharing the same borders? Is it possible to transform one into the other by flips? Regarding this we have:

THEOREM 7.6 *Given two triangulations of a set of sites on the cylinder, with the same borders, it is possible to transform each into the other by a sequence of flips.*

Proof: Before proving the result introduce two new concepts: we say that a convex vertex v in an upper (resp. lower) polygon could be: (a) an *Euclidean convex vertex*, when it and its adjacent vertices are in Euclidean position on the cylinder; or (b) an *essential convex vertex*, otherwise, see Figure 7.36.

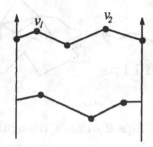

Figure 7.36. In the upper polygonal, v_1 is an Euclidean convex vertex, and v_2 is an essential convex vertex.

Now we proceed by induction on the number N of vertices of the set of sites. If N is less or equal to 6 it is easy to see 'by hand', that the result holds. Suppose that we have two triangulations T_1 and T_2 of a set of $N > 6$ sites on the cylinder, with the same borders and call U and L the upper, and respectively lower, polygons defining these borders. We have two possibilities:

1. There exists at least one Euclidean convex vertex v either in U or in L. If v is an ear in both triangulations just cut this ear and apply induction hypothesis. If v is an ear in T_1 but not in T_2 we proceed as follows. Consider the Euclidean polygon in the cylinder defined by v and its adjacent vertices in T_2. This is a triangulation of an Euclidean polygon, and using Theorem 7.5 we know that it is possible to transform it into a triangulation with an ear in v. So transform T_2 to such a triangulation T_2^*, and we have now a triangulation with an ear common with T_1, see Figure 7.37.

   If v is an ear in neither T_1 nor T_2, proceed as before transforming T_1 into a triangulation T_1^* with an ear in v, and T_2 into T_2^* with an ear in v; then apply the induction hypothesis.

2. There exists no Euclidean convex vertex in either U or L. In this case let v be an essential convex vertex (it is possible that no such vertex exists, but in this case U and L will be great circles, and we proceed by choosing as v any vertex of U or in L). Let N_1 and N_2 be the sets of vertices adjacent to v in T_1 and T_2, respectively. Consider $N = N_1 \cup N_2$ and sort this set of points circularly from v, and draw an Euclidean polygon P by just following this circular order. If there

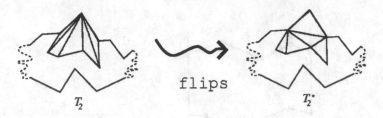

T_2 flips T_2^*

Figure 7.37. Transforming by flips T_2 into a triangulation with an ear in v.

exist some vertices in the interior of the Euclidean polygon the set of vertices of which is N add these vertices to N, re-sort circularly from v, and redraw P, see Figure 7.38.

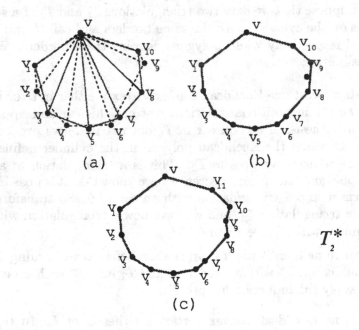

Figure 7.38. (a) Consider all vertices adjacent to v in both triangulations; (b) Sort these vertices circularly from v and draw an Euclidean polygon following this order; (c) 'Catch' in this Euclidean polygon any interior vertices not adjacent to v.

Now consider in P all original diagonals from T_1 and complete to a triangulation T_1^* of the set of sites; completing the original diagonals from T_2 obtain a triangulation T_2^*, Figure 7.39. It is clear that by removing P from T_1^* and T_2^* and applying the induction hypothesis that it is possible to transform T_1^* into T_2^* by a sequence of flips.

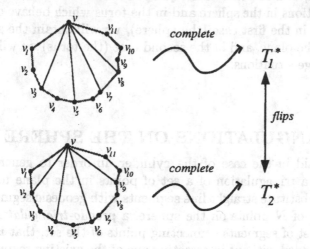

Figure 7.39. Constructing two new triangulations of the set of sites.

On the other hand, by just removing v and its adjacent vertices of T_1^* and T_1, and using the induction hypothesis, we have that we can transform each into the other by flips; and the same reasoning works for claiming that T_2 can be transformed into T_2^*, and we obtain the result in this case as well.

□

For essential polygons on the cylinder from Theorem 7.6 we have trivially:

COROLLARY 7.4 *Given two triangulations of an essential polygon on the cylinder it is possible to transform each into the other by a sequence of flips.*

3.   TRIANGULATIONS ON THE SPHERE AND ON THE TORUS

In the last section we have tried to show some of the characteristics of triangulations outside the plane, at least in the case of the cylinder. The behavior of those triangulations was sometimes very different from that in the plane, and in other cases quite similar. In this section we will see triangulations in the sphere and in the torus which behave quite differently, since, in the first case (the sphere), we will obtain the same properties as in the plane, and in the second case (the torus) we will see much more strange situations.

3.1   TRIANGULATIONS ON THE SPHERE

As we have said in the case of the cylinder, in order to generalize the definition of a triangulation of a set of points in the plane to the sphere, we will substitute straight line segments with geodesic segments. So given a set P of N points on the sphere, a *pseudo-triangulation of P* is a maximal set of segments connecting points of the set, that is, no segment can be added without intersecting one of the existing segments. In [Cortés et al., 1998] it is proved that any spherical polygon with more than four vertices admits a diagonal, and so pseudo-triangulations in the sphere are properly triangulations.

Given the importance of this structure, triangulations in the sphere have been considered in many occasions previously in different contexts (see, for instance [Brown, 1980, Nielson and Ramaraj, 1987, Okabe et al., 1992]). In paricular, we have mentioned Brown's work before who proved that Delaunay triangulations on the sphere can be obtained from the spatial convex hull of the set of sites embedded in Euclidean space. This is a first difference between triangulation on the sphere with respect to triangulations in the cylinder. In fact, the 'bad behavior' in many aspects observed in the previous section of the triangulations on the cylinder is not shared by the triangulations on the sphere, and these are, in some sense, closer to the plane than to the cylinder. Thus in [Cortés et al., 1998] it is proved that not only any spherical polygon with more than four vertices admits a diagonal and so the graph of triangulations of a polygon on the sphere is connected, but that even the graph of triangulations of a set of sites is connected (in fact, the proof of this result runs parallel to the planar case, since it is proved that any triangulation of a set of sites can be transformed into the Delaunay triangulation via diagonal flips).

3.2   TRIANGULATIONS ON THE TORUS

We have commented at the beginning of this section that triangulations on the sphere have a 'good behavior' compared to on the cylinder; in [Cortés et al., 1998] it is proved that triangulations on the torus can have the worst possible behavior.

Thus in Figure 7.40 there is shown a maximal set of segments for a given set of sites on the torus such that all faces are not triangles (in fact, one of its faces is not homeomorphic to a 2-cell).

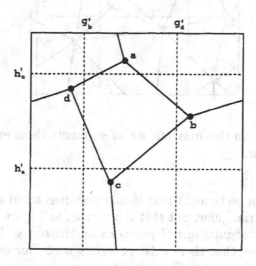

Figure 7.40. A maximal set of segments for a given set of sites on the torus such that all faces are non-triangular faces.

So, opposite to what happened on the cylinder or on the sphere, pseudo-triangulations and triangulations do not agree on the torus.

Notice that in Figure 7.40 a quadrant of the torus has no points; in fact, this property can be extended in the following way:

PROPOSITION 7.1 *[Cortés et al., 1998] Let S be a set of sites in the torus. If there exists no point of S in a quadrant of the torus then for any maximal set of segments in S there exists at least one non-triangular face.*

For some sets of sites there exist maximal sets of segments such that all their faces are triangles and other maximal sets of segments that do not verify this property, as Figures 7.41 and 7.42 show.

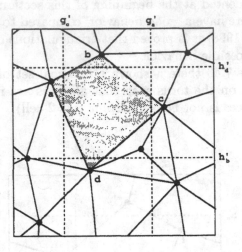

Figure 7.41.   In this maximal set of segments there exists a face that is not triangular.

Finally, it can be thought that in any maximal set of sites most of the faces must be triangular, but this is not true, as Figure 7.43 shows.

Regarding triangulations of polygons on the torus. Firstly, it is important to notice that there exist polygons with four or more vertices that admit no diagonal as Figure 7.44 shows.

Therefore, no pseudo-triangulation on the torus is a triangulation, and the graph of triangulations of a polygon on the torus can be empty. Moreover, as happened with a set of sites, even if the polygon admits triangulations a greedy algorithm does not necessarily lead to one of them (see Figure 7.45).

THE GRAPH OF TRIANGULATIONS OF A TOROIDAL POLYGON

Note that all known proofs of the connectivity of the graph of triangulations (either in the plane or in the cylinder) are based, in some way, on the property that a given polygon is always triangulable, but this is not true in the case of the torus, so if the graph of triangulations of a

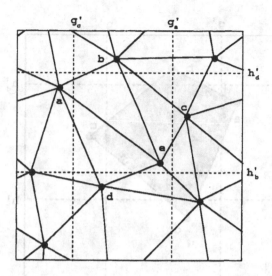

Figure 7.42. A new maximal set of segments for the same set of sites as in Figure 7.41, but now all faces are triangular.

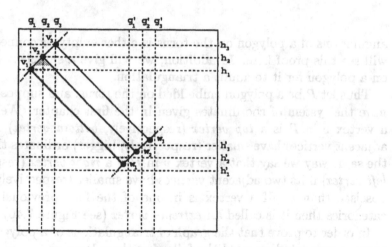

Figure 7.43. It is not possible to bound the number of non-triangular faces in a maximal set of segments.

polygon on the torus is connected a new kind of proof must be found. In fact, it has been proved in [Cortés et al., 2001] that the graph of tri-

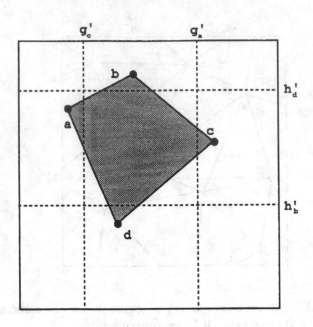

Figure 7.44.   This polygon has no diagonal.

angulations of a polygon on the torus is either empty or connected. We will see this proof here. In addition, we will give a sufficient condition on a polygon for it to admit a triangulation.

Thus let P be a polygon embedded on the torus, and suppose that we have the system of coordinates given in the first chapter. We say that a vertex v in P is a *top vertex* (respectively, *bottom vertex*) if its two adjacent vertices have smaller (respectively, bigger) ordinates than v. In the same way we say that a vertex v in P is a *right vertex* (respectively, *left vertex*) if its two adjacent vertices have smaller (respectively, bigger) abscisae than v. If a vertex is in one of the four previously defined categories then it is called an *extreme vertex* (see Figure 7.46).

In order to prove that the graph of triangulations of a polygon on the torus is connected we need the following three lemmas.

LEMMA 7.7 *Any triangulable quadrilateral on the torus admits a triangulation that has an ear in one of its extreme vertices.*

Proof: We only need to prove the lemma for the case of two extreme vertices (otherwise the lemma is trivial). Those vertices can be either adjacent or not. If they are adjacent one of the two ears must be an

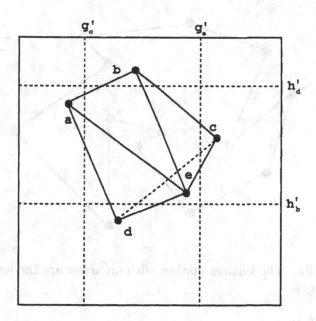

Figure 7.45. This polygon with five vertices admits a triangulation, but if we start a greedy algorithm with the diagonal *cd* we obtain the polygon of Figure 7.44 which cannot be triangulated.

extreme vertex, then we can assume that they are not adjacent. This situation leads to one of the two last cases of Figure 7.47, in the case b) at least one of the ears is an extreme vertex. But in the case c) the original triangulation can have no ears in the extreme vertices, but in this case it is possible to flip the internal diagonal so that the result holds.

□

Now we extend the result given in Lemma 7.7 to any polygon.

LEMMA 7.8 *Any triangulable polygon on the torus admits a triangulation that has an ear in one of its extreme vertices.*

Proof: Let *P* denote a polygon with a given triangulation *T*; we are going to rebuild *P* incrementally from four of its vertices and using for this construction the given triangulation defined in *P*. By Lemma 7.7 the quadrilateral P_4 formed by the four initial vertices has at least one ear that is an extreme vertex. Now we add to P_4 a new vertex *v* that

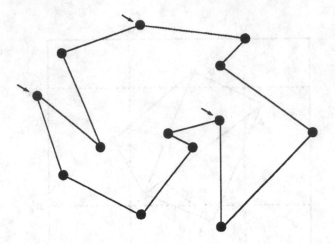

Figure 7.46. The vertices marked with an arrow are the top vertices of this polygon.

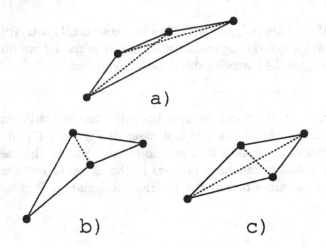

Figure 7.47. The three possible quadrilaterals.

must be adjacent to two vertices of P_4. If the new vertex is not adjacent to the ear of P_4 that is an extreme vertex the new polygon P_5 will verify the enunciation of the lemma. Otherwise the old ear will disappear, so we must assure that $P-5$ admits a triangulation with an ear in one of its extreme vertices.

Consider the region R exterior to P_4 delimited by the meridian and the parallel which pass through the vertices of P_4 adjacent to v, as Figure 7.48 shows. If v is outside that region then v is an ear of that triangulation that is an extreme vertex (Figure 7.48 a)). If, on the contrary, v is in the interior of R then we can check that it is possible to flip the diagonal defined by the two vertices adjacent to v (one of the them is the former ear) and in the new triangulation the former ear is again an ear which is an extreme vertex.

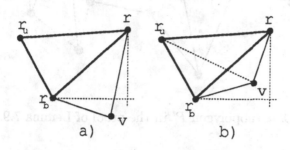

Figure 7.48.   Adding a new vertex to a polygon.

We can repeat the process adding a new vertex each time, and the cases will be the same as those previously studied for five points.   □

Now we can prove the key result in order to obtain the connectivity of the graph of triangulations of a polygon on the torus.

LEMMA 7.9 *Let P be a polygon on the torus such that there exists a triangulation with an ear which is an extreme vertex u, then any triangulation of P can be transformed into other with an ear in v by a sequence of flips.*

Proof: For the sake of simplicity we can assume that u is a top vertex. Let T be a new triangulation of P, then we have to prove that it is possible to transform T into a new triangulation T' such that one of the ears of T' is just u.

Now let P' be the subpolygon of P defined by all vertices which are adjacent to u in T (see Figure 7.49). We divide the vertices of $P' - \{u\}$ into two subsets V_l and V_r containing the vertices of P' which are on the left and on the right of u respectively.

We label the vertices of V_l counter-clockwise starting at the vertex adjacent to u in the polygon. If v_1 'sees' v_3 (i.e., the diagonal $v_1 v_3$ is interior to the polygon) then we flip the edge $v_1 v_2$ to $v_1 v_3$ and then we

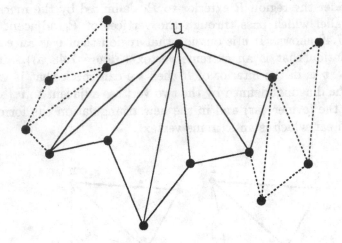

Figure 7.49.   The subpolygon P' in the proof of Lemma 7.9.

delete v_2 from V_l (and from P'). Otherwise we continue the process in the same way, obtaining, in the end, a new set V_l' such that all its vertices other than the first and the last are reflex vertices (see Figure 7.50). Carrying out the same process with V_r we obtain a polygon as that depicted in Figure 7.51.

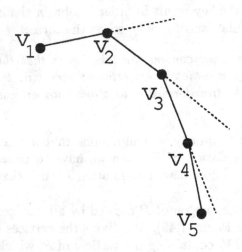

Figure 7.50.   The set V_l' (see the proof of Lemma 7.9).

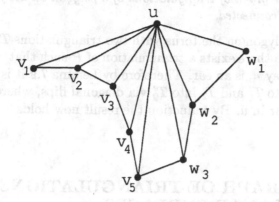

Figure 7.51. The polygon obtained in the intermediate step of the proof of Lemma 7.9.

If we now call v_s the last vertex of V_l and w_t the last vertex of V_r (both are adjacent) we will concentrate in the pentagon $uv_{s-1}v_sw_tw_{t-1}$. In that pentagon it is always possible to flip one diagonal (see Figure 7.52, and thus making a new flip we can transform the original triangulation into a new triangulation such that u is an ear in that new triangulation. So the result holds.

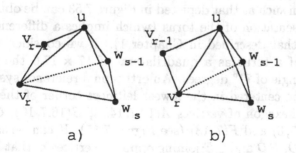

Figure 7.52. The existence of the top vertex u allows always a diagonal in a pentagon to be flipped.

□

Therefore we are in a position to enunciate the main result of this section.

THEOREM 7.7 *The graph of triangulations of a polygon on the flat torus is either empty or connected.*

Proof: If P is a polygon on the torus with two triangulations T_1 and T_2 then by Lemma 7.8 there exists a triangulation of P such that one of its extreme vertices, say u, is an ear. Therefore by Lemma 7.9 it is possible to transform T_1 into T_1' and T_2 into T_2' via diagonal flips, where T_1' and T_2' have both an ear in u. By induction the result now holds.   □

4.   THE GRAPH OF TRIANGULATIONS ON NON-PLANAR SURFACES

After the results on the graph of triangulations obtained on the cylinder, the sphere, and the torus, one can think that this graph is connected for all polygons on all surfaces. In fact, the opposite is almost true, because all surfaces (including the sphere and the torus) admit a metric such that some polygons have non-connected graphs of triangulations. The results of this section appear in [Cortés et al., 2001].

The key to the result of this section is obtained by a simple observation. If H is a hexagon all diagonal except exactly those connecting opposite vertices (see Figure 7.53), then the graph of triangulations of H has two elements and no flip is possible; therefore that graph of triangulations is non-connected (see Figure 7.54).

And a hexagon such as that depicted in Figure 7.53 can be obtained in a suitable representation of the torus (which implies a different metric in the torus of that described in Chapter 1). If we consider a planar representation of a torus as a quadrilateral of 17×17.6 the sides of which form an angle of $71°$ degrees. An orthogonal reference system can be assumed to be centered in the lowest leftmost corner of the square. We can draw a hexagon of vertices $A(10.7, 14.5)$, $B(16.7, 10)$, $C(15, 2)$, $D(10.5, 2.5)$, $E(5, 6)$ and $F(7, 13)$ (see Figure 7.55). It can be seen that the diagonals AD, BD and CE joining opposite vertices in that hexagon are non-admissible (they are not segments). So two triangulations can be found such as no flips in any of them are possible.

In addition, any other orientable surface can be obtained from this representation of the torus just by adding handles in one corner of the fundamental region on the Figure 7.55 as Figure 7.56 shows, therefore, any orientable surface admits a metric such that the graph of triangulations of some polygons in those surfaces with that metric are non-connected. Even more there exist metrics on the sphere that admit a hexagon such as that shown in Figure 7.55.

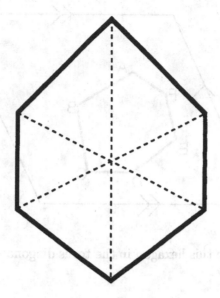

Figure 7.53. The dashed diagonals are declared non-admissible in this hexagon.

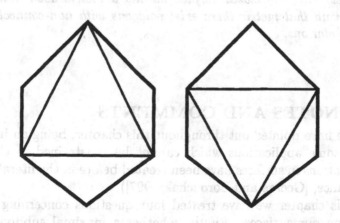

Figure 7.54. The two triangulations of the hexagon of Figure 7.53.

Regarding non-orientable surfaces, first of all we have to notice that a hexagon with the same set of non-admissible diagonals can be obtained

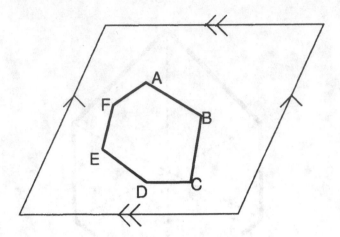

Figure 7.55. In this hexagon in the torus diagonals AD, BE and CF are non-admissible.

in the projective plane. Thus as any other non-orientable surface can be obtained from the projective plane just by gluing Möbius bands. Summarizing we have obtained:

THEOREM 7.8 *Any closed surface admits a metric such that on that surface with that metric there exist polygons with non-connected graph of triangulations.*

5.   NOTES AND COMMENTS

As we have pointed out throughout this chapter, being an important tool in some applications which cannot be constrained to the plane, triangulations in surfaces have been treated before in the literature (see, for instance, [George and Borouchaki, 997]).

In this chapter we have treated four questions concerning triangulations on our surfaces. Firstly, whether a maximal subdivision is a triangular subdivision; we have seen that both concepts are equivalent in the case of the cylinder or the sphere but not in the case of the torus (on this last surface it is possible to obtain maximal subdivisions such that all faces are not triangular). Secondly, what is the domain defined by a set of sites when we perform a triangulation. We have seen that if the points are in non-Euclidean position this domain is not unique in

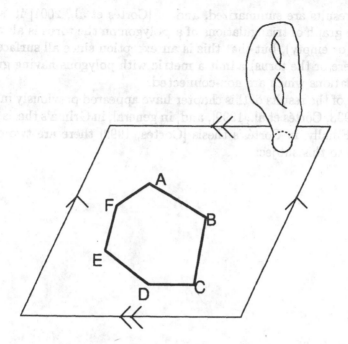

Figure 7.56. Adding handles to the torus obtained in Figure 7.55 it is possible to obtain the same hexagon such that diagonals AD, BE and CF are non-admissible.

the case of the cylinder, and it is defined by an upper polygonal and a lower polygonal. In the case of the sphere that domain is always the m-convex hull. Nevertheless, it is not clear how that domain is on the torus, this being one important open problem.

The third question is whether it is possible to go from a triangulation to another by using diagonal flips? In the plane this has been a very useful method for obtaining optimal triangulations, and it is known that this method can be applied without problem on the sphere; but on the cylinder, although the answer to this third question is 'yes', we have seen that it cannot be applied for obtaining optimal triangulations, this happens because it is not possible to identify a 'target' triangulation as the Delaunay triangulation in the plane. Nevertheless, an incremental randomized algorithm leads to a triangulation which maximizes the minimum angle, obtaining in this way an answer to the last question.

Regarding the torus, things are quite a bit more complicated, on the one hand, maximal subdivisions do not agree with triangular subdivisions, and there exists polygons that do not admit a triangulation, the known results are summarized, and in [Cortés et al., 2001] it is proved that the graph of triangulations of a polygon on the torus is always connected (or empty), but that this is an exception since all surfaces (even the sphere or the torus) admit a metric with polygons having graphs of triangulations which are non-connected.

Some of the results of this chapter have appeared previously in [Cortés et al., 1998, Cortés et al., 1999], and, in general, in Grima's thesis [Grima, 1998]. Finally, in Cortés's thesis [Cortés, 1999] there are two chapters devoted to this subject.

References

[Agarwal and Sharir, 1993] Agarwal, P. and Sharir, M. (1993). Circular visibility of a simple polygon from a fixed point. *Internat. J. Comput. Geom. Appl.*, 3:1–25.

[Aho et al., 1974] Aho, A., Hopcroft, J., and Ullman, J. (1974). *The design and analysis of computer algorithms*. Addison-Wesley, Reading, Mass.

[Akl, 1979] Akl, S. (1979). Two remarks on a convex hull algorithm. *Information Processing Letters*, 8:108–109.

[Akl and Toussaint, 1978] Akl, S. and Toussaint, G. (1978). Efficient convex hull algorithms for pattern recognition applications. In *Proc. 4th. Int'l Joint Conf. on Pattern Recognition*, pages 483–487, Kyoto, Japan.

[Amato et al., 1994] Amato, N. M., Goodrich, M. T., and Ramos, E. A. (1994). Parallel algorithms for higher-dimensional convex hulls. In *Proc. 35th Annu. IEEE Sympos. Found. Comput. Sci.*, pages 683–694.

[Ash and Bolker, 1985] Ash, P. and Bolker, E. (1985). Recognizing dirichlet tessellations. *Geometriae Dedicata*, 19:175–206.

[Atallah and Bajaj, 1987] Atallah, M. and Bajaj, C. (1987). Efficient algorithms for common transversals. *Inform. Process. Lett.*, 25:87–91.

[Aurenhammer, 1991] Aurenhammer, F. (1991). Voronoi diagrams: A survey of a fundamental data structure. *ACM Comput. Surveys*, 23:345–405.

[Avis and Wenger, 1988] Avis, D. and Wenger, R. (1988). Polyhedral line transversals in space. *Discrete and Comput. Geometry*, 3:257–266.

[B. Grünbaum, 1956] B. Grünbaum (1956). A proof of Vazsonyi's conjecture. *Bull. Res. Council Israel*, A(6):77–78.

[Bartllet, 1975] Bartllet, M. (1975). *The Statistical Analysis of Spatial Pattern*. London: Chapman and Hall.

[Bielawski, 1987] Bielawski, R. (1987). Simplicial convexity and its applications. *J. Math. Anal. Appl.*, 127:155–171.

[Blanc, 1943] Blanc, E. (1943). Les ensembles surconvexes plans. *Ann. Sci. École Norm. Sup.*, 60:463–468.

[Bogue, 1949] Bogue, D. (1949). The structure of the metropolitan community: A study of dominance and subdominance. *Ann Arbor: Horace M. Rackham School of Graduate Studies, University of Michigan.*

[Boissonnat and Yvinec, 1998] Boissonnat, J. and Yvinec, M. (1998). *Algorithmic Geometry*. Cambridge University Press.

[Brown, 1965] Brown, G. (1965). Point density in stems per acre. *New Zealand Forestry Service Research Notes*, 38:1–11.

[Brown, 1980] Brown, K. (1980). *Geometric transformations for fast geometric algorithms*. PhD thesis, Dept. of Computer Science, Carnegie Mellon Univ.

[Busemann and Phadke, 1979] Busemann, H. and Phadke, B. (1979). Minkowskian geometry, convexity conditions and the parallel axiom. *J. Geom.*, 12:17–33.

[Cavendish, 1974] Cavendish, J. (1974). Automatic triangulation of arbitrary planar domains for the finite element method. *Int. J. Numerical Methods in Engineering*, 8:679–696.

[Chan, 1995] Chan, T. (1995). Output-sensitive results on convex hulls, extreme points, and related problems. In *11th Annu. ACM Sympos. Comput. Geom.*, pages 10–19. ACM.

[Chan et al., 1995] Chan, T., Snoeyink, J., and Yap, C. (1995). Output-sensitive construction of polytopes in four dimensions and clipped voronoi diagrams in three. In *6th ACM-SIAM Sympos. Discrete Algorithms (SODA'95)*, pages 282–291. ACM-SIAM.

[Chazelle et al., 1993] Chazelle, B., Edelsbrunner, H., Guibas, L., and Sharir, M. (1993). Diameter, width, closest line pair and parametric searching. *Discrete Comp. Geom.*, 10:183–196.

[Cobos et al., 1997a] Cobos, F., Dana, J., Grima, C., and Márquez, A. (1997a). Anchura de un conjunto convexo en la esfera. In *Proc. VII Encuentros en Geometría Computacional*. Universidad Politécnica de Madrid.

[Cobos et al., 1997b] Cobos, F., Dana, J., Grima, C., and Márquez, A. (1997b). Diameter of a set on the cylinder. In *Proc. Ninth Canadian Conference on Computational Geometry*, Kingston (Canada).

[Cobos et al., 1997c] Cobos, F., Dana, J., Grima, C., and Márquez, A. (1997c). Diámetro de un conjunto de puntos en el cilindro. In *Proc. VII Encuentros en Geometría Computacional*. Universidad Politécnica de Madrid.

[Cobos et al., 1997d] Cobos, F., Dana, J., Grima, C., and Márquez, A. (1997d). The width of a convex set on th sphere. In *Proc. Ninth Canadian Conference on Computational Geometry*, Kingston (Canada).

[Cobos et al., 1995] Cobos, F., Dana, J., Marquez, A., and Mateos, F. (1995). Helices en el cilindro y en el toro y planificacion de movimientos de robots. In Hurtado, F. and Sacristan, V., editors, *VI Encuentro de Geometria Computacional*, pages 332–339. UPC.

[Cortés, 1999] Cortés, C. (1999). *Triangulaciones en superficies*. PhD thesis, Universidad de Sevilla.

[Cortés et al., 2001] Cortés, C., Grima, C., Hurtado, Márquez, A., Santos, F., and Valenzuela, J. (2001). Transforming triangulations of polygons on non-planar surfaces. Preprint.

[Cortés et al., 1999] Cortés, C., Grima, C., and Márquez, A. (1999). Flipping Edges in Triangulations on the Cylinder. In Bronnimann, H., editor, *Proc. 15th European Workshop on Comp. Geom.* INRIA.

[Cortés et al., 1998] Cortés, C., Hurtado, F., Márquez, A., and Nakamoto, A. (1998). Edge flipping in triangulations of surfaces. *14th European Workshop on Computational Geometry*, Universidad Politécnica de Cataluña(Barcelona).

[Dacey, 1965] Dacey, M. (1965). The geometry of central place theory. *Geografiska Annaler*, 47B:111–124.

[Dana et al., 1995] Dana, J., , Garrido, M., and Marquez, A. (1995). Visibilidad con obstaculos en el cilindro. In Hurtado, F. and Sacristan, V., editors, *VI Encuentro de Geometria Computacional*, pages 125–132. UPC.

[Dana et al., 1998] Dana, J., Grima, C., Hurtado, F., and Márquez, A. (1998). More results on the computational geometry on the cylinder. In *Proc. 14th European Workshop on Computational Geometry*. Universidad Politécnica de Cataluña.

[Dana et al., 1997] Dana, J., Grima, C., and Márquez, A. (1997). Convex hull in non-planar surfaces. In *Proc. 13th European Workshop on Computational Geometry*, Germany. University of Wuerzburg.

[Danzer et al., 1963] Danzer, L., Grunbaum, B., and Klee, V. (1963). Helly's theorem and its relatives. In *Proc. Symp. Pure Math. VII*, Providence, R.I. American Mathematical Society.

[de Berg et al., 1997] de Berg, M., van Kreveld, M., Overmars, M., and Schwarzkkopf, O. (1997). *Computational Geometry: algorithms and applications*. Springer-Verlag.

[Dehne and Klein, 1987] Dehne, F. and Klein, R. (1987). An optimal algorithm for computing the Voronoi diagram on a cone. Report SCS-TR-122, School of Computer Science, Carleton University, Ottawa, Canada.

[Dirichlet, 1850] Dirichlet, G. (1850). Über die reduction der positeven quadratischen formen mit drei unbestimmten ganzen zahlen. *Journal für die Reine und Angewandte Mathematik*, 40:209–227.

[do Carmo, 1976] do Carmo, M. (1976). *Geometría diferencial de curvas y superficies*. Alianza Universidad Textos.

[do Carmo, 1992] do Carmo, M. (1992). *Riemannian Geometry*. Birkhauser Boston, Inc., Boston, MA.

[Edelsbrunner, 1985] Edelsbrunner, H. (1985). Finding transversals for sets of simple geometric figures. *Theoret. Comput. Sci.*, 35:55–69.

[Edelsbrunner, 1987] Edelsbrunner, H. (1987). *Algorithms in Combinatorial Geometry*. Springer-Verlag.

[Edelsbrunner et al., 1982] Edelsbrunner, H., Maurer, A., Preparata, F., Rosenberg, A., Welzl, E., and Wood, D. (1982). Stabbing line segments. *BIT*, 22:274–281.

[Erdos, 1946] Erdos, P. (1946). On sets of distances of n points. *Amer. Meth. Monthly*, 53:248–250.

[Field, 1986] Field, D. (1986). Implementing Watson's algorithm in three dimensions. In *Proc. 2nd Annu. Symp. Comp. Geom.*, pages 246–259. ACM.

[Freeeman, 1974] Freeeman, H. (1974). Computer processing of line-drawing images. *Comput. Surveys*, 6:57–97.

[Freeman, 1974] Freeman, H. (1974). Computer processing of line-drawing images. *Comput. Surveys*, (6):57–97.

[Freeman and Shapira, 1975] Freeman, H. and Shapira, R. (1975). Determining the minimum-area encasing rectangle for an arbitrary closed curve. *Comm. ACM*, 18(7):409–413.

[George, 1981] George, J. (1981). Computer implementation of the finite element method. Tech. Rep. STAN-CS-81-208. Computer Scince Dept. Stanford University.

[George and Borouchaki, 997] George, P. and Borouchaki, H. (!997). *Triangulation de Delaunay et Maillage*. Ed. Hermes, Paris.

[Getis and Boots, 1978] Getis, A. and Boots, B. (1978). *Models of Spatial Processes: An Approach to the Study of Point, Line and Area Patterns*. Cambridge University Press.

[Goldberg, 1969] Goldberg, M. (1969). A solution of problem 66-11: Moving furniture through a hallway. *SIAM Review*, 11:75–78.

[Goodman and O'Rourke, 1997] Goodman, J. and O'Rourke, J. (1997). *Handbook of Discrete and Computational Geometry*. CRC Press.

[Graham, 1972] Graham, R. (1972). An efficient algorithm for determining the convex hull of a finite planar set. *Inform. Process. Lett.*, 1:132–133.

[Grima, 1998] Grima, C. (1998). *Geometría Computacional en superficies*. PhD thesis, Universidad de Sevilla.

[Grima et al., 1998a] Grima, C. I., Márquez, A., and Ortega, L. (1998a). A locus approach to angle problems in computational geometry. In *14th European Workshop in Computational Geometry*, Barcelona.

[Grima et al., 1998b] Grima, C. I., Márquez, A., and Ortega, L. (1998b). A locus approach to angle problems in computational geometry. In *Wet and Discrete*, Darwin (Australia).

[Gritzmann and Klee, 1993] Gritzmann, P. and Klee, V. (1993). Computational complexity of inner and outer j-radii of polytopes in finite dimensional normed spaces. *Math. Programming*, 59:163–213.

[Gruber, 1993] Gruber, P. (1993). History of convexity. In *Handbook of convex Geometry*. Elsevier Science.

[Guibas et al., 1992] Guibas, L., Knuth, D., and Sharir, M. (1992). Randomized incremental construction of Delaunay and Voronoi diagrams. *Algorithmica*, 7:381–413.

[Harborth, 1974] Harborth, H. (1974). Solution to problem 664a. *Elem. Math.*, 29:14–15.

[Harding, 1921] Harding, J. (1921). Calculation of ore tonnage and grade from drill-hole samples. *Transactions of the American Institute of Mining Engineers*, 66:117–126.

[Harding, 1923] Harding, J. (1923). How to calculate tonnage and grade of an ore body. *Engineering and Mining Journal Press*, 116(11):445–448.

[Hershberger and Suri, 1996] Hershberger, J. and Suri, S. (1996). Offline maintenance of planar configurations. *Journal of Algorithms*, 21:453–475.

[Hocking and Young, 1961] Hocking, J. and Young, G. (1961). *Topology*. Addison-Wesley, Reading, MA.

[Hopf and Rinow, 1931] Hopf, H. and Rinow, W. (1931). Über den Begriff der vollständigen differential-geometrischen Fläche. *Comentarii. Math. Helvetici.*, 3:209–225.

[Horton, 1917] Horton, R. E. (1917). Rational study of rainfall data makes possible better estimates of water field. *Engineering News-Record*, 79:211–213.

[Houle and Toussaint, 1988] Houle, M. and Toussaint, G. (1988). Computing the width of a set. *IEEE Trans. on pattern analysis and machine intelligence*, 10(5):761–765.

[Howden, 1968] Howden, W. (1968). The sofa problem. *Computer Journal*, 11:299–301.

[Hurtado, 1993] Hurtado, F. (1993). *Problemas Geométricos de Visibilidad*. PhD thesis, Universitat Politècnica de Catalunya.

[Hurtado and Noy, 1996] Hurtado, F. and Noy, M. (1996). Tiangulations, visibility graphs and reflex vertices of a simple polygon. *Computational Geometry: Theory and its Applications*, 6:355–369.

[Hurtado et al., 2000] Hurtado, F., Sacristán, V., and Toussaint, G. (2000). Some constrained minimax and maximin location problems. *Studies in Locational Analysis*, 15:17–35.

[Ichida and Kiyono, 1975] Ichida, K. and Kiyono, T. (1975). Segmentation of plane curves. *Trans. Elec. Commun. Eng. Japan*, 58-D:689–696.

[Imai and Iri, 1988] Imai, H. and Iri, M. (1988). Polygonal approximation of a curve: Formulations and solution algorithms. In Toussaint, G. T., editor, *Computational Morphology*. North–Holland, Amsterdam, The Netherlands.

[Jaromczyk and Kowaluk, 1988] Jaromczyk, J. and Kowaluk, M. (1988). Skewed projections with an application to line stabbing in r^3. In *Proc. 4thAnn. ACM Symp. Comput. Geom.*, pages 362–370.

[Katchalski et al., 1985] Katchalski, M., Lewis, T., and Zaks, J. (1985). Geometric permutations for convex sets. *Discrete Math*, 54:271–284.

[Kendall and Moran, 1963] Kendall, M. and Moran, P. (1963). *Geometrical Probability*. Hafner, New York.

[Kirkpatrick and Seidel, 1986] Kirkpatrick, D. and Seidel, R. (1986). The ultimate planar convex hull algorithm? *SIAM J. Comput.*, 15:287–299.

[Klein, 1988] Klein, R. (1988). Abstract voronoi diagrams and their applications. In *Lecture Notes in Computer Science*, volume 333, pages 148–157, Berlin. Springer-Verlag.

[Klein, 1989] Klein, R. (1989). *Concrete and Abstract Voronoi Diagrams*. Springer-Verlag.

[Klein and Wood, 1988] Klein, R. and Wood, D. (1988). Voronoi diagrams based on general metrics in the plane. In *Lecture Notes in Computer Science*, volume 294, pages 281–291, Berlin. Springer-Verlag.

[Knuth, 1973] Knuth, D. (1973). *Sorting and Searching: The Art of Computer Programming III*. Addison-Wesley, Reading,MA.

[Knuth, 1976] Knuth, D. (1976). Big omicron and big omega and big theta. *SIGACT News*,, 8(2):18–24.

[Kurozumi and Davis, 1982] Kurozumi, Y. and Davis, W. (1982). Polygonal approximation by the minimax method. *Comput. Graphics Image Processing*, 19:248–264.

[Lawson, 1972] Lawson, C. (1972). Transforming triangulations. *Discrete Math.*, 3:365–372.

[Lee and Wu, 1986] Lee, D. and Wu, Y. (1986). Geometric complexity of some location problems. *Algorithmica*, 1:193–211.

[Mani-Levitska, 1993] Mani-Levitska, P. (1993). Characterizations of convex sets. In *Handbook of convex Geometry*. Elsevier Science.

[Márquez and Valenzuela, 2000] Márquez, A. and Valenzuela, J. (2000). Computing the Voronoi diagram of a set of segments on the cylinder.

[Maruyama, 1973] Maruyama, K. (1973). An approximation method for solving the sofa problem. *International Journal of Computer and Information Sciences*, 2:29–48.

[Mazón, 1992] Mazón, M. (1992). *Diagramas de Voronoi en caleidoscopios*. PhD thesis, Universidad de Cantabria.

[Mazón and Recio, 1997] Mazón, M. and Recio, T. (1997). Voronoi diagrams on orbifolds. *Journal Computational Geometry: Theory and Applications*, 8(5):219–230.

[Megiddo, 1983] Megiddo, N. (1983). Linear-time algorithms for linear programming in $\mathbf{R}^3$ and related problems. *SIAM J. Comput.*, 12:759–776.

[Meister, 1975] Meister, G. (1975). Polygons have ears. *Amer. Math. Mon.*, 82:648–651.

[Menger, 1928] Menger, K. (1928). Urtersuchungen über allgemeine Metrik. *Math. Ann.*, (100):75–163.

[Miles, 1971] Miles, R. (1971). Random points, sets and tessellations on the surface of a sphere. *Sankhya: The Indian Jounal of Statistics, Series A*, 33(2):145–174.

[Moser, 1966] Moser, L. (1966). Problem 66-11: Moving furniture through a hallway. *SIAM Review*, 8:381.

[Moser and Pach, 1993] Moser, W. and Pach, J. (1993). Recent developments in combinatorial geometry. In Pach, J., editor, *New Trends in Discrete and Computational Geometry*. Springer-Verlag, New York.

[Nielson and Ramaraj, 1987] Nielson, M. and Ramaraj, R. (1987). Interpolation over a sphere based upon a minimum norm network. *Computetr Aided Geometric Design*, 4:41–57.

[Niggli, 1927] Niggli, R. (1927). Die topologische strukturanalyse. *Zeitschrift für Kristallographie*, 65:391–415.

[Nikulin and Shafarevich, 1987] Nikulin, V. and Shafarevich, I. (1987). *Geometries and Groups*. Springer, Berlin.

[Nowacki, 1976] Nowacki, V. (1976). Über allgemeine eigenschaften von wirkungsbereichen. *Zeitschrift für Kristallographie*, 143:360–385.

[Okabe et al., 1992] Okabe, A., Boots, B., and Sugihara, K. (1992). *Spatial Tessellations: Concepts and Applications of Voronoi Diagrams*. John Wiley and Sons.

[O'Rourke, 1981] O'Rourke, J. (1981). An on-line algorithm for fitting straight lines between data ranges. *Comm. ACM.*, 24:574–578.

[O'Rourke., 1987] O'Rourke., J. (1987). *Art Gallery Theorems and Algorithms*. Oxford University Press.

[O'Rourke, 1994] O'Rourke, J. (1994). *Computacional Geometry in C.* Cambridge University Press.

[Paschinger, 1982] Paschinger, I. (1982). *Konvexe Polytope und Dirichletsche Zellenkomplexe*. PhD thesis, Institut für Mathematik, University of Salzburg, Austria.

[Pei, 1984] Pei, L. (1984). Convex fuzzy sets i. *J. Wuhan Univ. Natur. Sci. Ed.*, pages 13–22.

[Pielou, 1977] Pielou, E. (1977). *Mathematical Ecology*. New York: Wiley-Interscience.

[Popoff, 1966] Popoff, C. (1966). Computing reserves of mineral deposits: principles and conventional methods. *U.S. Department of the Interior, Bureau of Mines, Information Circular, 8283*.

[Preparata and Shamos, 1985] Preparata, F. and Shamos, M. (1985). *Computational geometry: an introduction*. Springer-Verlag, New York.

[Rényi and Sulanke, 1963] Rényi, A. and Sulanke, R. (1963). Ueber die konvexe Hulle von n zufalling gewahlten Punkten, I. *Z. Wahrschein*, 2:75–84.

[Rosenfeld, 1969] Rosenfeld, A. (1969). *Picture Processing by Computers*. Academic Press, New York.

[Sack and Urrutia, 2000] Sack, J. and Urrutia, J. (2000). *Handbook of Computational Geometry*. Elsevier Science, The Netherlands.

[Sacristán, 1997] Sacristán, V. (1997). *Optimización Geométrica y Aplicaciones en Visibilidad*. PhD thesis, Universidad Politécnica de Cataluña.

[Schaudt and Drysdale, 1991] Schaudt, B. and Drysdale, R. (1991). Multiplicatively weighted crystal growth voronoi diagrams. *Proc. 7th Annu. Symp. Comp. Geom. ACM*, pages 214–223.

[Schwartz et al., 1987] Schwartz, J., Sharir, M., and Hopcroft, J. (1987). *Algorithmic and Geometric Robotics*. Lawrence Earlbaum Assoc.

[Schwartz and Yap, 1987] Schwartz, J. and Yap, C. (1987). *Planning Geometry and Complexity of Robot Motion*. Ablex Publishing.

[Sebastian, 1970] Sebastian, J. (1970). A solution to problem 66-11: Moving furniture through a hallway. *SIAM Review*, 12:582–586.

[Seidel, 1997] Seidel, R. (1997). Convex hull computations. In *Handbook of Discrete and Computational Geometry*, pages 361–375. J.E. Goodman and J. O'Rourke, CRC Press.

[Shamos, 1978] Shamos, M. (1978). *Computational geometry*. PhD thesis, Dept. Comput. Sci., Yale Univ.

[Shieh, 1985] Shieh, Y. (1985). Rau and the economic law of market areas. *Journal of Regional Science*, 25(2):191–199.

[Sklansky, 1972] Sklansky, J. (1972). Measuring concavity on a rectangular mosaic. *IEEE Trans. Comp.*, C-21:1355–1364.

[Snyder, 1962] Snyder, D. (1962). Urban places in uruguay and the concept of a hierarchy. *Festschrift: C.F. Jones, Northwestern University, Studies in Geography*, 6:29–46.

[Strang, 1982] Strang, G. (1982). The width of a chair. *The American Mathematical Monthly*, 89(8):529–534.

[Strang and Fix, 1973] Strang, G. and Fix, G. (1973). *An Analysis of the Finite Element Method*. Prentice-Hall, Englewood Cliffs, NJ.

[Takeda, 1985] Takeda, S. (1985). *On geographical optimization and dynamic facility problem*. PhD thesis, University of Tokyo.

[Thiessen, 1911] Thiessen, A. (1911). Precipitation averages for large areas. *Monthly Weather Review*, 39:1082–1084.

[Toussaint, 1985] Toussaint, G. (1985). Movable separability of sets. In Toussaint, G. T., editor, *Computational Geometry*. Elsevier Science.

[Upton and Fingleton, 1985] Upton, G. and Fingleton, B. (1985). *Spatial Data Analysis by Example. Volume 1: Point Pattern and Quantitative Data Chichester*. John Wiley.

[V. Icke and R. Van de Weygaert, 1987] V. Icke and R. Van de Weygaert (1987). Fragmenting the universe. i statistics of two-dimensional voronoi foams. *Astronomy and Astrophysics*, 184:16–32.

[Voronoi, 1908] Voronoi, G. (1908). Nouvelles applications des paramètres continus à la théorie des formes quadratiques, deuxième memoire, recherches sur les sur les parallelloèdres primitifs. *Journal für die Reine und Angewandte Mathematik*, 134:198–287.

[Wendel, 1962] Wendel, J. (1962). A problem in geometric probability. *Math. Scand.*, 2:109–111.

[Whitney, 1929] Whitney, E. (1929). Areal rainfall estimates. *Monthly Weather Review*, 57:462–463.

[Wigner and Seitz, 1933] Wigner, E. and Seitz, F. (1933). On the constitution of metallic sodium. *Physical Review*, 43:804–810.

Topic Index

185

Author Index

189